Circular Economy and Sustainability

This book series aims at exploring the rising field of Circular Economy (CE) which is rapidly gaining interest and merit from scholars, decision makers and practitioners as the global economic model to decouple economic growth and development from the consumption of finite natural resources. This field suggests that global sustainability can be achieved by adopting a set of CE principles and strategies such as design out waste, systems thinking, adoption of nature-based approaches, shift to renewable energy and materials, reclaim, retain, and restore the health of ecosystems, return recovered biological resources to the biosphere, remanufacture products or components, among others.

However, the increasing complexity of sustainability challenges has made traditional engineering, business models, economics and existing social approaches unable to successfully adopt such principles and strategies. In fact, the CE field is often viewed as a simple evolution of the concept of sustainability or as a revisiting of an old discussion on recycling and reuse of waste materials. However, a modern perception of CE at different levels (micro, meso, and macro) indicates that CE is rather a systemic tool to achieve sustainability and a new eco-effective approach of returning and maintaining waste in the production processes by closing the loop of materials. In this frame, CE and sustainability can be seen as a multidimensional concept based on a variety of scientific disciplines (e.g., engineering, economics, environmental sciences, social sciences). Nevertheless, the interconnections and synergies among the scientific disciplines have been rarely investigated in depth.

One significant goal of the book series is to study and highlight the growing theoretical links of CE and sustainability at different scales and levels, to investigate the synergies between the two concepts and to analyze and present its realization through strategies, policies, business models, entrepreneurship, financial instruments and technologies. Thus, the book series provides a new platform for CE and sustainability research and case studies and relevant scientific discussion towards new system-wide solutions.

Specific topics that fall within the scope of the series include, but are not limited to, studies that investigate the systemic, integrated approach of CE and sustainability across different levels and its expression and realization in different disciplines and fields such as business models, economics, consumer services and behaviour, the Internet of Things, product design, sustainable consumption & production, bio-economy, environmental accounting, industrial ecology, industrial symbiosis, resource recovery, ecosystem services, circular water economy, circular cities, nature-based solutions, waste management, renewable energy, circular materials, life cycle assessment, strong sustainability, and environmental education, among others.

Srikanta Patnaik • Leon Miller
Francesco Vigliarolo
Editors

Circular Economy and Sustainable Value Creation through Eco-Innovation

Springer

Editors
Srikanta Patnaik
Bhubaneswar, Odisha, India

Francesco Vigliarolo
Bologna, Italy

Leon Miller
Tallinn University of Technology
Tallinn, Estonia

ISSN 2731-5509 ISSN 2731-5517 (electronic)
Circular Economy and Sustainability
ISBN 978-3-031-88515-0 ISBN 978-3-031-88516-7 (eBook)
https://doi.org/10.1007/978-3-031-88516-7

This Springer imprint is published by the registered company Springer Nature Switzerland AG
The registered company address is: Gewerbestrasse 11, 6330 Cham, Switzerland

Preface

Since the time of Adam Smith, economics has undergone a significant transformation toward positivism. Economic science has increasingly focused on studying systems "as they are" using mathematical language, rather than projecting what they should achieve based on social forecasts or aspirations. Neville Keynes, the father of the renowned economist John Maynard Keynes, played a pivotal role in this evolution. In his work, *Studies and Exercises in Formal Logic* (1884), he emphasized clarity and mathematical symbolism. Later, he distinguished between positive economics—the description of how an economic system functions—and normative economics—the evaluation of its desirability in terms of costs and benefits. This shift led economics to focus on models of infinite monetary growth driven by individual interests, largely omitting essential questions such as: What kind of society do we want? Consequently, this classical and neoclassical notion of infinite economic growth profoundly influenced the concept of development, initially equating it with national economic growth post-World War II.

The economic development paradigm emerged in the post-colonial era, centered on improving the economic performance of underdeveloped nations with a primary focus on GDP. However, the paradigm, shaped by international experts employing top-down planning, failed to incorporate local values, cultural heritage, and aspirations. Its inadequacies became apparent: it ignored bottom-up inclusiveness, failed to balance economic growth with human and social development, and caused extensive environmental degradation and climate issues.

By the late twentieth century, the limitations of this paradigm gave rise to the sustainability concept—a more balanced, inclusive, and holistic approach to improving human well-being, economic performance, social values, and environmental stewardship. Sustainability promotes an integrative approach to creating a better future, unifying economic, social, and environmental dimensions. This approach enhances prosperity, happiness, and public value, while addressing environmental challenges. It emphasizes co-creating knowledge and employing advanced technologies, ensuring societal goals align with what people value most.

Sustainability planning has since evolved to include the circular economy, which integrates resource reuse and disposal in harmony with natural rhythms. The circular economy promotes principles, such as waste elimination, renewable energy adoption, ecosystem restoration, and product remanufacturing. These strategies align with the broader goals of the UN's Sustainability Agenda, blending technology, eco-innovation, and human aspirations to achieve harmonious development. The focus on sustainability involves planning for future growth in a way that safeguards the natural environment and the atmosphere and avoids depleting natural resources. However, the term sustainability also has broader implications that include co-creating a quality of life that is not disrupted for any environmental, economic, or social reasons. Therefore, this edited compilation also addresses the significance of foresight and integrity in co-creating a circular economy. By adding the circular concept, sustainability planning includes circular streams of resource use and disposal—this means that future planning is in harmony with natural rhythms and patterns, which are circular rather than linear. In terms of planning for growth, the circular economy calls for using resources in a way in which they can be returned to nature without any disruption to natural patterns and rhythms. In the circle economy, there is no waste; thus, social agents act in the same way as nature, which means everything is completely recycled. This causes planners and theorists to think in terms of biodegradability, reuse, or refurbishment. Therefore, theorists and practitioners agree that the goals of the Sustainability Agenda can be achieved by adopting circular economy principles and strategies, e.g., design out waste, systems thinking, adoption of nature-based approaches, shift to renewable energy and materials, reclaim, retain, and restore the health of ecosystems, return recovered biological resources to the biosphere, and remanufacture products or components, among others.

Thus, in addition to analyzing the extent to which technological innovations are in line with sustainability and the circular economy, the contributors also address issues related to futurology, foresight, integrity, and co-creation. The integrity concept is part of a larger social movement toward a more holistic, integrative, and balanced approach to future planning. Foresight is a term used in conceptualizations of the sustainability concept to describe a critical analysis of the impact that new technology, artificial intelligence, and robotics have on achieving sustainability goals (with the sustainability goals including establishing a circular economy). Foresight also involves planning how eco-innovations and advances in technology can be integrated into future sustainability planning in a way that enables humanity to enjoy the best of all possible futures. Integrity is a perspective on how the future is envisioned, conceptualized, and planned. Integral futures, in this respect, mean an approach to sustainability that is inclusive of enhancing the eco-aesthetic dimensions of socio-cultural life, planning that is inclusive of human and social development (which includes increasing human and social well-being), and integrity also implies increasing prosperity and flourishing. Co-creation means that social stakeholders interact in value creation networks that generate the knowledge of how to produce outcomes that enable them to live in accordance with what they have reason to value, enjoy a higher quality of life, and with fellow stakeholders co-create what they regard as the best of all possible futures.

Practitioners and theorists are increasingly interested in the integrative approach to co-creating a sustainable future and, as a means, effectively making the types of social and value transformations necessary for a futuristic circle economy. The broader integrative perspective is a value transformation that involves interrelational and public–private processes for co-creating future value. Therefore, a desired social value transformation requires an improvement in the performance of individuals, private organizations, public institutions, and entire social economies. The outcome is human, social, and sustainable development, which means not just improving the conditions of society, improving the economy, or successfully adjusting to Industry 4.0, but elevating the human experience. In this respect, this book offers various theoretical and practical explanations regarding what it means to grow as individuals, organizations, institutions, and societies. The integrative approach is people-centered in that it places people and their growth and development first. However, this approach is also based on the realization that the desired transformation of organizations, institutions, and social economies requires developing the capabilities of the individuals working in those various capacities.

Indeed, stakeholders interacting at the multiple levels of social activity (from local to international) increasingly call for a re-evaluation and transformation of the basic assumptions and values that seem to have placed us on a path that is non-sustainable. This collection of articles represents the commitment of the editors, conjointly with Springer Publishing, to provide cutting-edge insight and state-of-the-art knowledge on sustainability, the circular economy, and various aspects of human, social, and development planning that are related to these core topics. Therefore, the edited book, *Circular Economy and Sustainable Value Creation Through Eco-Innovation*, offers a collection of perspectives on how to move together toward a future that balances human, social, and sustainable development. Because the collection contains the knowledge and experience of scholars and practitioners from diverse backgrounds around the world, it combines theory and practice to provide insight that is relevant for various stakeholders who are concerned with successful value transformations. A successful value transformation is shift away from less favorable human conditions toward a more desired state of human, social, and global existence. Thus, one significant aim of this edited book is to highlight the growing theoretical links between human, social, and sustainable development. Therefore, this book investigates the synergies between the five concepts (i.e., sustainability, circular economy, foresight, co-creation, and integrity) and analyzes the prospect of integrating these five concepts in order to create strategies, policies, and models for improving the quality of life in multiple dimensions. Furthermore, it proposes concepts such as reciprocity and meso-economy, socially oriented economies, and innovative approaches to interpret economic systems in terms of social meaning, such as a phenomenological approach, in order to understand how the circular economy is an important tool for promoting the demand for people's rights. Thus, this edited book provides a platform for an integrative approach to sustainability practice and research plus relevant discussion toward new system-wide solutions to our current problems.

Bhubaneswar, Odisha, India — Prof. (Dr.) Srikanta Patnaik
Tallinn, Estonia — Prof. (Dr.) Leon Miller
Bologna, Italy — Prof. (Dr.) Francesco Vigliarolo

Acknowledgement

We would like to express our sincere appreciation and gratitude to all those who have contributed their research findings in this edited volume *Circular Economy and Sustainable Value Creation Through Eco-Innovation*. Their active involvement, insightful discussions, and valuable contributions made the book an asset of value.

Our heartfelt appreciation goes out to our esteemed editors Prof. Srikanta Patnaik, Prof. Leon Miller, and Prof. Francesco Vigliarolo for their valuable insights and perspectives which are instrumental for this edited volume. We are grateful for the time and effort they took to make this volume a true success.

Our deepest appreciation goes out to the reviewer committee for their diligent work in reviewing and selecting the chapters that had got place in this volume. Their expertise and commitment to academic excellence helped to ensure that the chapters were of highest quality.

We are grateful to all the individuals and organizations that played a part in the success of the edited volume on *Circular Economy and Sustainable Value Creation Through Eco-Innovation*. Without your support and dedication, this volume would not have been possible. Thank you all for your contributions and commitment to advancing the field of computational finance and business analytics.

Srikanta Patnaik
Leon Miller
Francesco Vigliarolo

Editorial

In the rapidly changing global landscape, the traditional linear economic model—focused on the "take, make, dispose" philosophy—has reached its limits in terms of sustainability. The growing pressures of environmental degradation, resource depletion, and climate change have led to the rise of a more responsible and resilient approach: the circular economy. In this book, we delve deeply into the multi-faceted transformation of business and industry as they transition from linear models to circular approaches, emphasizing the significance of sustainable value creation and eco-innovation as key pillars of this shift.

The Paradigm Shift from Linear Economy Model to Circular Economy

In previous decades, economies have followed a linear model where products were designed, manufactured, used, and ultimately discarded. This approach, while successful in fueling industrial growth, has become unsustainable, with significant environmental costs. The circular economy, however, provides an alternative that focuses on extending the lifecycle of products, reducing waste, and regenerating resources. By embracing principles of reusing, recycling, and refurbishing, the circular model strives to minimize waste while maximizing the use of available materials. Our exploration throughout the chapters highlights how companies are adopting circular business models that prioritize sustainability at every stage of the value chain. This shift is not only crucial for mitigating environmental harm but also opens up new avenues for innovation and growth.

Sustainable Value Creation: Beyond Profit Maximization

The concept of sustainable value creation extends far beyond the traditional idea of profit maximization. While profit is undoubtedly important, it is no longer the sole determinant of business success. The modern business landscape demands that organizations consider the long-term impacts of their actions on both society and the environment. Sustainable value creation focuses on producing goods and services that meet present needs without compromising the ability of future generations to meet their own needs. In our chapters, we explore how businesses are integrating environmental and social factors into their value creation models, with a particular emphasis on the triple bottom line—people, planet, and profit. By embedding sustainability into core business practices, companies can build long-term resilience, enhance brand reputation, and cultivate consumer loyalty.

Eco-Innovation: Driving Change and Opportunity

At the heart of this transformation is eco-innovation, the process of developing new products, services, or processes that not only drive economic growth but also contribute to environmental sustainability. Eco-innovation represents an opportunity to redesign industries in ways that reduce their ecological footprint while creating new market opportunities. In our book, we explore the mechanisms of eco-innovation—how companies are leveraging cutting-edge technologies, sustainable design practices, and innovative business models to provide a more sustainable future. From green technologies to waste reduction systems, eco-innovation drives the creation of new solutions that meet both consumer demand and environmental needs. It is an essential tool in achieving the goals of the circular economy and ensuring that businesses remain competitive in a market increasingly focused on sustainability.

The integration of circular economy principles, sustainable value creation, and eco-innovation is more than just a business strategy—it is a social imperative. By adopting these practices, society stands to benefit in numerous ways. First and foremost, the reduction of waste and resource consumption directly contributes to environmental preservation, mitigating issues like pollution and climate change. Additionally, the focus on sustainable practices encourages green jobs, stimulates local economies, and fosters a more equitable distribution of resources. The shift to circular business models also promotes a more collaborative economy, where industries, governments, and communities work together to solve pressing global challenges. Ultimately, these principles ensure that future generations inherit a planet capable of supporting their needs, while also promoting economic stability and prosperity.

In this book, we have explored the integral relationship between the circular economy, sustainable value creation, and eco-innovation, showing how businesses are leading the charge in redefining what it means to succeed in the modern economy. As industries move from linear to circular models, they not only redefine their operations but also transform their relationship with the environment and society. Sustainable value creation and eco-innovation are not mere buzzwords but essential components of a future where business success is measured by its positive contribution to the planet and its people. This paradigm shift offers immense opportunities for innovation, growth, and societal well-being, and it is our hope that this book inspires both thought leaders and business practitioners to embrace these principles and lead the charge toward a more sustainable future.

Contents

About the Editors

Srikanta Patnaik is the Director of IIMT, Bhubaneswar. He received his Ph.D. (Engineering) on Computational Intelligence from Jadavpur University, India, in 1999, and he has supervised 34 Ph.D. theses and more than 60 M.Tech theses in the areas of machine intelligence, soft computing applications, and re-engineering. Dr. Patnaik has published more than 100 research papers in international journals and conference proceedings. He is the author of 2 textbooks and has edited 82 books and few invited book chapters, published by leading international publishers, such as Springer-Verlag and Kluwer Academic. He is the Editor-in-Chief of the *International Journal of Information and Communication Technology* and the *International Journal of Computational Vision and Robotics* published by Inderscience Publishing House, England. He also serves as the Editor-in-Chief of the book series "Modeling and Optimization in Science and Technology," published by Springer, Germany.

Leon Miler completed his PhD at Tallinn University of Technology, Estonia, specializing in "The Theory and Practice of Value Creation." In addition, he completed doctor-level studies in social development, peace research, and consulting at the University of Chicago's Cluster of Theological Schools. He is the founder and coordinator of "A Value-Based Approach to Sustainability." He is also a research fellow affiliated with the Network of Religious and Traditional Peacemakers and Tallinn University of Technology. He is an instructor of ethics, intercultural relations, intercultural communications, and organizational behavior. He has experience working with The International Association of Religious Freedom promoting Human Rights and improved interethnic relations as well as experience analyzing the Cyprus conflict (the UN's longest running peacekeeping mission) in cooperation with the European Peace Research Association. He is currently involved in researching human, social, and sustainable development with academic partners in South Asia.

Francesco Vigliarolo holds a PhD in Social Sciences, specializing in economic sociology, ethical finance, and social economics. He is a full-time Professor of Regional Economy at the Catholic University of La Plata. Additionally, he also serves as the Director of the UNESCO Chair in "Economic Systems and Human Rights" at the National University of La Plata. He is also an Associate Researcher at the CESOT (Center for the Study of Sociology of Work) at the University of Buenos Aires. His primary focus lies in local development processes, particularly the construction of territorial identities that involve asserting rights (ontological reason) through associative forms of production and ethical finance tools. Throughout his career, he had published numerous books and scientific articles delving into concepts, such as "ethical social capital," "economic socialization and democratization, "the principles of economic phenomenology," "ontological reason of economy and demand for rights of economy," and "meso-economic approach."

Chapter 1
The Circular Economy: A Paradigm Shift Towards Sustainability with Reference to an Indian City Raipur

Deepti Patnaik and Srikanta Patnaik

Abstract The circular economy is emerging as a transformative solution to sustainability challenges, particularly in rapidly urbanizing cities like Raipur, India. This study explores the principles and applications of the circular economy, emphasizing waste reduction, resource conservation, and economic resilience. Raipur, as an industrial hub, faces significant environmental challenges, including waste mismanagement, resource depletion, and pollution. By adopting circular economy strategies—such as recycling, green infrastructure, sustainable construction, and community-driven waste management—the city can transition from a linear, wasteful economic model to a regenerative and resource-efficient system. Case studies highlight successful initiatives, including waste management programs, eco-friendly building projects, and innovative recycling start-ups. The study underscores the role of policy interventions, financial incentives, and public engagement in fostering circular practices. Implementing extended producer responsibility (EPR), green procurement policies, and educational campaigns can accelerate this transition. While challenges such as infrastructure gaps, regulatory enforcement, and public awareness persist, strategic collaboration among government, businesses, and communities can drive sustainable development. This research contributes to the discourse on circular economy adoption in emerging economies, demonstrating its potential to balance economic growth with environmental stewardship.

Keywords Circular economy · Sustainability · Waste management · Resource efficiency · Urban development

D. Patnaik (✉)
Faculty of Commerce & Management, Kalinga University, Naya Raipur, Chhattisgarh, India
e-mail: deepti.patnaik@kalingauniversity.ac.in

S. Patnaik
Interscience Institute of Management & Technology, Bhubaneswar, Odisha, India

S. Patnaik et al. (eds.), *Circular Economy and Sustainable Value Creation through Eco-Innovation*, Circular Economy and Sustainability,
https://doi.org/10.1007/978-3-031-88516-7_1

1.1 Introduction

The circular economy represents a profound and transformative shift in economic and environmental thinking, positioning itself as a sustainable alternative to the traditional linear model of economic growth. At its core, the circular economy is defined as an economic system aimed at eliminating waste through the continuous use of resources. This model emphasizes designing products and processes in such a way that waste is minimized, and materials are kept in circulation within the economy for as long as possible (Ellen MacArthur Foundation, 2013). Unlike the conventional linear economy, which operates on a 'take, make, dispose' model, the circular economy promotes a regenerative approach that seeks to close the loop of product lifecycles through practices such as recycling, reusing, and remanufacturing.

The importance of sustainability in the modern economy cannot be overstated. As global populations grow and industrial activities expand, the pressure on natural resources and ecosystems intensifies. Traditional economic models have led to significant environmental challenges, including resource depletion, pollution, and climate change. The Brundtland Commission's seminal report, *Our Common Future*, highlights that sustainability involves meeting the needs of the present without compromising the ability of future generations to meet their own needs (Brundtland Commission, 1987). This concept encompasses not only environmental protection but also economic and social dimensions. Sustainable development aims to balance economic growth with environmental stewardship and social equity, ensuring that economic activities do not deplete natural resources or degrade the environment.

The circular economy is integral to the modern sustainability agenda, as it addresses the limitations of the linear model by promoting resource efficiency and minimizing environmental impact. By designing out waste and pollution, keeping products and materials in use, and regenerating natural systems, the circular economy aligns with broader sustainability goals. This model encourages a shift from a throwaway culture to one where products are designed for longevity, repairability, and recyclability. The Ellen MacArthur Foundation (2013) outlines that the circular economy offers a pathway to reducing environmental footprint, conserving resources, and fostering economic resilience through innovative business practices and technologies.

Raipur, the capital city of Chhattisgarh in India, serves as a pertinent case study for examining the application of circular economy principles. As an emerging urban center with a rapidly growing population and expanding industrial base, Raipur faces significant challenges related to waste management, resource scarcity, and environmental degradation. The city's industrial activities, including steel manufacturing, cement production, and agriculture, are resource-intensive and generate substantial waste. According to the Central Pollution Control Board (CPCB, 2021), Raipur produces approximately 600 tonnes of municipal solid waste daily, much of which is inadequately managed, leading to environmental pollution and health hazards.

In this context, the circular economy offers a strategic framework for addressing environmental and economic challenges. By integrating circular economy principles into local policies and practices, the state can enhance waste management, conserve resources, and improve environmental quality. For instance, promoting eco-friendly building materials and reducing single-use plastics can mitigate environmental impact and resource depletion. Establishing community recycling programs, repair cafes, and second-hand markets can extend the life cycle of products and reduce waste. Moreover, investing in green infrastructure, such as urban parks and community gardens, can contribute to regenerating natural systems and improving public health.

The relevance of the circular economy to state extends beyond environmental benefits. It presents opportunities for economic development and social advancement by fostering new business models and creating job opportunities in sectors such as recycling, repair, and green manufacturing. For example, circular economy practices can drive innovation in product design and manufacturing, leading to the development of new technologies and services that align with sustainability goals. Additionally, community-based initiatives, such as recycling programs and repair workshops, can strengthen local economies and enhance social cohesion by engaging residents in sustainable practices.

The state's journey towards embracing the circular economy is reflective of broader trends in urban sustainability and economic transformation. The city's efforts to integrate circular economy principles align with global movements towards sustainable urbanization and resource management. As it continues to grow and develop, the adoption of circular economy practices will be crucial in achieving long-term resilience and prosperity. The city's experience can serve as a model for other urban centers in India and beyond, demonstrating the potential of circular economy approaches in addressing complex environmental and economic challenges.

The circular economy represents a paradigm shift towards sustainability, offering a comprehensive approach to addressing the limitations of the traditional linear model. For Raipur, the integration of circular economy principles provides a pathway to tackle environmental challenges, promote economic development, and enhance social well-being. As the city navigates its path towards sustainable development, the adoption of circular economy practices will play a pivotal role in shaping a resilient and prosperous future. The circular economy's potential to transform economic practices and environmental outcomes underscores its significance in the modern economy and its relevance to cities like Raipur.

1.2 Conceptual Framework

1.2.1 Principles of the Circular Economy

The circular economy is underpinned by three fundamental principles designed to create a sustainable economic system that contrasts sharply with the traditional linear economy. These principles are: design out waste and pollution, keep products and materials in use, and regenerate natural systems.

1.2.1.1 Design Out Waste and Pollution

Designing out waste and pollution involves rethinking how products and processes are designed and implemented to minimize environmental impact. This principle emphasizes the importance of incorporating sustainability into the design phase, ensuring that products are created with their entire lifecycle in mind. By focusing on reducing waste and pollution from the outset, the circular economy seeks to prevent environmental degradation and resource depletion.

In practical terms, this principle can be applied by selecting materials that are durable, non-toxic, and recyclable. For instance, designers might use materials that are both renewable and biodegradable, thus reducing the potential for pollution. Additionally, designing products for easy disassembly ensures that components can be easily repaired, refurbished, or recycled, further minimizing waste (Ellen MacArthur Foundation, 2013).

1.2.1.2 Keep Products and Materials in Use

Keeping products and materials in use emphasizes the importance of extending the lifecycle of products through reuse, repair, refurbishment, and recycling. This principle aims to maximize the value derived from products and materials by keeping them in circulation for as long as possible before they reach the end of their useful life.

This approach can be seen in various practices such as developing business models that facilitate product leasing or sharing, establishing repair and refurbishment centers, and creating systems for efficient recycling. For example, a company might offer a product-as-a-service model where customers pay for the service provided by the product rather than owning it outright. This encourages the company to design products for longevity and ease of maintenance, knowing that the product will return to them at the end of its life for refurbishment or recycling (Ellen MacArthur Foundation, 2013).

1.2.1.3 Regenerate Natural Systems

Regenerating natural systems focuses on restoring and enhancing natural ecosystems that have been degraded by human activities. This principle acknowledges the need to go beyond sustainability by actively contributing to the health and regeneration of natural environments.

In practice, this might involve implementing agricultural practices that restore soil fertility and biodiversity, reforestation projects that rebuild forest ecosystems, or urban greening initiatives that enhance city landscapes. For example, creating green spaces in urban areas not only improves air quality and provides recreational spaces but also supports urban biodiversity and helps mitigate the heat island effect (Ellen MacArthur Foundation, 2013).

1.2.2 Differences Between Circular and Linear Economies

The traditional linear economy and the circular economy represent two contrasting approaches to economic activity and resource management.

1.2.2.1 Resource Flow and Waste Management

In a linear economy, the flow of resources follows a 'take, make, dispose' model. Raw materials are extracted, used to create products, and then discarded as waste at the end of their lifecycle. This model often leads to significant resource depletion and environmental pollution, as products are designed for a single use and are not easily recyclable or biodegradable (Ellen MacArthur Foundation, 2013).

In contrast, the circular economy aims to create a closed-loop system where resources are continuously cycled through the economy. Products are designed to be durable, repairable, and recyclable, with the goal of minimizing waste and maximizing resource efficiency. Waste from one process becomes input for another, reducing the need for new raw materials and lessening environmental impact.

1.2.2.2 Economic Value and Business Models

The linear economy typically focuses on maximizing short-term profits through the sale of products and services, often at the expense of long-term sustainability. Business models in a linear economy prioritize high-volume production and consumption, with little regard for the environmental costs associated with waste and resource depletion (Ellen MacArthur Foundation, 2013).

The circular economy, on the other hand, encourages the development of business models that create long-term value by emphasizing product longevity, resource efficiency, and environmental stewardship. These models often include product-as-a-service, take-back schemes, and closed-loop supply chains. By focusing on extending product lifecycles and minimizing waste, circular business models aim to generate economic value while contributing to environmental and social sustainability.

1.2.2.3 Environmental Impact

The linear economy is associated with high levels of environmental degradation due to its reliance on finite resources and the production of large amounts of waste. This approach contributes to problems such as resource depletion, pollution, and climate change (Ellen MacArthur Foundation, 2013).

The circular economy seeks to mitigate these environmental impacts by designing systems that reduce waste and pollution, promote resource efficiency, and regenerate natural systems. By closing the loop of product lifecycles and emphasizing sustainable practices, the circular economy aims to create a more resilient and less environmentally damaging economic system.

1.2.2.4 Consumer Behavior

In the linear economy, consumer behavior is typically oriented towards ownership and consumption. Products are often viewed as disposable, leading to frequent replacements and significant waste (Ellen MacArthur Foundation, 2013).

The circular economy encourages a shift in consumer behavior towards valuing durability, repairability, and recyclability. It promotes practices such as product leasing, sharing, and second-hand consumption. This shift supports a more sustainable approach to consumption, where products are used for longer periods and their lifecycle is managed more effectively.

The principles of the circular economy—designing out waste and pollution, keeping products and materials in use, and regenerating natural systems—represent a comprehensive approach to creating a more sustainable and resilient economic system as shown in Fig. 1.1. These principles contrast sharply with the traditional linear economy, which is characterized by a one-way flow of resources and a focus on short-term profit at the expense of long-term sustainability. By adopting circular economy principles, businesses, governments, and individuals can contribute to a more sustainable future, addressing the environmental and economic challenges of the modern world.

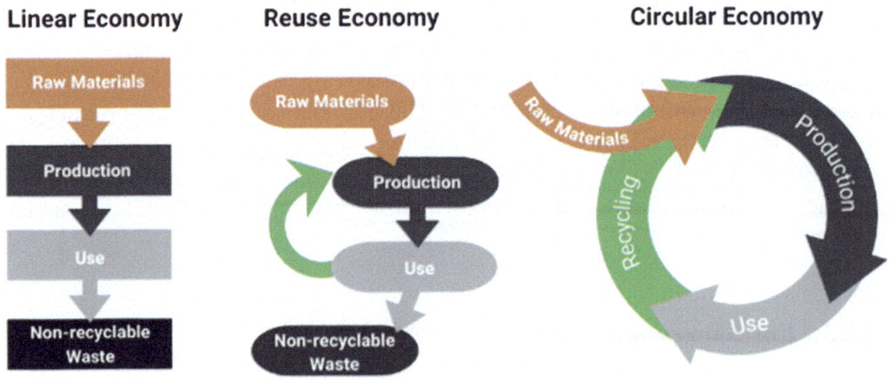

Fig. 1.1 Pictorial representation of the Linear, Reuse and Circular Economy

1.3 The Need for a Circular Economy

Raipur, the capital city of Chhattisgarh in India, is rapidly evolving as a significant urban center with a burgeoning population and an expanding industrial base. This growth, while contributing to economic development, has also led to a range of pressing economic and environmental challenges. Addressing these challenges through the principles of a circular economy offers a pathway to sustainability and resilience for the city. This section explores Raipur's current economic and environmental issues, including waste management problems, resource scarcity, and environmental degradation, and outlines the need for a circular economy to address these concerns effectively.

1.3.1 Waste Management Issues

Raipur's waste management system is under significant strain due to the city's rapid urbanization and population growth. According to the Central Pollution Control Board (CPCB), Raipur generates approximately 600 tonnes of municipal solid waste (MSW) daily, a figure that has been steadily increasing with the rise in population and economic activity (CPCB, 2021). The waste management infrastructure in the city faces several challenges:

1. **Inadequate Infrastructure**: The existing waste management facilities, including collection, transportation, and processing systems, are insufficient to handle the growing volume of waste. Many areas in Raipur lack effective waste segregation and collection services, leading to uncollected waste accumulating in public spaces and illegal dumpsites.
2. **High Proportion of Organic Waste**: A significant portion of the waste generated is organic, including food and garden waste. The absence of effective composting or anaerobic digestion facilities means that much of this organic waste ends up in landfills, where it contributes to methane emissions and other environmental issues.
3. **Low Recycling Rates**: Despite efforts to improve recycling, the rates remain low. Many recyclable materials are not separated at the source or are contaminated, making them less viable for recycling. This results in valuable resources being lost to landfills and incineration.

These waste management challenges underscore the need for a circular economy approach in Raipur, which can improve waste reduction, enhance recycling efforts, and support the development of effective waste processing infrastructure.

1.3.2 Resource Scarcity

Resource scarcity is another critical issue facing Raipur, driven by the city's industrial activities and increasing consumption patterns. Several factors contribute to this problem:

1. **Industrial Demands**: Raipur's economy is heavily dependent on industries such as steel manufacturing, cement production, and mining. These industries are resource-intensive and place significant pressure on local resources, including water and minerals. For instance, steel production requires large quantities of iron ore and coal, while cement production demands substantial amounts of limestone.
2. **Water Scarcity**: The city faces challenges related to water scarcity, exacerbated by industrial water consumption and inadequate water management practices. The depletion of local water sources, coupled with irregular monsoon patterns, has led to concerns about the sustainability of water supply for both domestic and industrial use.
3. **Depleting Natural Resources**: The extraction of natural resources for industrial purposes contributes to resource depletion and environmental degradation. Overexploitation of resources, coupled with inadequate conservation efforts, threatens the long-term availability of essential materials.

The circular economy can address resource scarcity by promoting resource efficiency, encouraging the use of recycled materials, and reducing dependency on virgin resources. By adopting circular principles, Raipur can enhance resource conservation and support the development of more sustainable industrial practices.

1.3.3 Environmental Degradation

Environmental degradation in Raipur is a significant concern, driven by the city's industrial activities, urbanization, and inadequate environmental management practices. Key issues include:

1. **Air Pollution**: Industrial activities, vehicle emissions, and construction dust contribute to high levels of air pollution in Raipur. The city often experiences elevated concentrations of particulate matter (PM) and other pollutants, which pose health risks to residents and impact overall air quality.
2. **Land Degradation**: Rapid urban expansion and industrial development have led to the conversion of green spaces and agricultural land into built-up areas. This land degradation reduces biodiversity, disrupts ecosystems, and contributes to the urban heat island effect.
3. **Soil Contamination**: Industrial activities, particularly mining and manufacturing, have resulted in soil contamination from hazardous waste and chemical run-

off. Contaminated soil can affect agricultural productivity and pose risks to public health.

4. **Water Pollution**: The discharge of untreated or inadequately treated industrial effluents and domestic sewage into water bodies has led to water pollution in Raipur. Polluted water sources can impact aquatic ecosystems and pose health risks to communities relying on these water sources for drinking and sanitation.

Addressing environmental degradation through the circular economy involves adopting practices that reduce pollution, enhance ecological restoration, and promote sustainable land and water management. By integrating circular principles into urban planning and industrial operations, Raipur can mitigate environmental impacts and improve overall ecological health.

1.4 Socio-Economic Context of Raipur

Raipur, the capital city of Chhattisgarh, India, is a dynamic urban center experiencing rapid socio-economic transformation. This transformation is driven by significant industrial growth and urbanization trends, which have shaped the city's economic landscape and created both opportunities and challenges. Understanding Raipur's industrial profile and urbanization trends is essential for assessing the need for a circular economy and developing strategies to address the city's environmental and resource management issues.

1.4.1 Industrial Profile

Raipur's industrial profile reflects a diverse and expanding economic base, characterized by significant growth in various sectors. The city has emerged as a key industrial hub in central India, driven by its strategic location, availability of resources, and investment in infrastructure. The major components of Raipur's industrial profile include:

Steel Industry: Raipur is a prominent center for steel production, with several large steel plants operating in the region. The city benefits from its proximity to rich iron ore deposits, which are essential for steel manufacturing. Major steel producers, such as Jindal Steel & Power Ltd., Bhushan Steel, and others, have established significant operations in Raipur, contributing to the city's economic growth and industrial prominence.

Cement Industry: The cement industry is another critical sector in Raipur's industrial landscape. The city's proximity to limestone deposits, a key raw material for cement production, has led to the establishment of several cement plants. Companies like Ambuja Cements and UltraTech Cement have significant manufacturing facilities in and around Raipur, driving economic development and employment in the region.

Mining and Mineral Processing: Raipur's industrial base also includes mining and mineral processing activities. The region is rich in various minerals, including coal, limestone, and bauxite. Mining operations contribute to the city's economic growth, but they also pose environmental challenges related to land degradation, water pollution, and resource depletion.

Textiles and Other Manufacturing: In addition to heavy industries, Raipur has a growing textile sector and other manufacturing industries. The city hosts several small and medium-sized enterprises (SMEs) involved in the production of textiles, garments, and consumer goods. These industries contribute to the city's economic diversity and provide employment opportunities.

Service Sector Growth: Alongside traditional industries, the service sector in Raipur has been expanding, driven by the growth of retail, education, healthcare, and real estate. The development of infrastructure, including shopping malls, educational institutions, and healthcare facilities, has bolstered the city's economic profile and improved the quality of life for residents.

While the industrial sector has been a major driver of economic growth in Raipur, it has also led to environmental challenges, including waste generation, pollution, and resource depletion. Addressing these challenges through the principles of a circular economy can enhance sustainability and resilience in the city's industrial activities.

1.4.2 Urbanization Trends

Raipur has experienced significant urbanization over the past few decades, reflecting broader trends of rapid urban growth in India. The key aspects of urbanization trends in Raipur include:

Population Growth: Raipur's population has grown substantially, driven by both natural population increase and migration from rural areas. According to the 2011 Census, the city had a population of approximately 1.1 million, and this figure has continued to rise. Rapid population growth has increased the demand for housing, infrastructure, and public services, placing additional pressure on urban resources and management systems (Census of India, 2011).

Infrastructure Development: The city has undergone extensive infrastructure development to support its growing population and economic activities. This includes the expansion of transportation networks, such as roads and railways, as well as improvements in utilities and public services. Urban infrastructure projects, including the development of residential, commercial, and industrial areas, have transformed the cityscape and facilitated economic growth.

Real Estate and Housing: The real estate sector in Raipur has witnessed significant expansion, with the construction of new residential and commercial properties. The growth of housing developments, including both affordable and luxury options, reflects the city's increasing urbanization and demand for real estate. However, rapid urban expansion has also led to challenges related to land use, green space preservation, and urban planning.

Environmental Impact of Urbanization: The rapid pace of urbanization has resulted in several environmental issues, including the loss of green spaces, increased air and water pollution, and pressure on waste management systems. The expansion of built-up areas has led to the reduction of natural habitats and agricultural land, contributing to environmental degradation and reduced biodiversity.

Social and Economic Disparities: Urbanization has also highlighted social and economic disparities within the city. While some areas have experienced rapid development and economic growth, others face challenges related to inadequate infrastructure, limited access to basic services, and socio-economic inequality. Addressing these disparities requires comprehensive urban planning and inclusive development strategies.

1.4.3 Implications for the Circular Economy

The socio-economic context of Raipur highlights the need for a circular economy approach to address the city's challenges and promote sustainable development. By integrating circular economy principles, Raipur can enhance resource efficiency, improve waste management, and mitigate environmental impacts associated with industrial and urban growth.

1. **Improving Waste Management**: Circular economy practices can help Raipur address its waste management issues by promoting recycling, composting, and waste-to-energy technologies. Effective waste management systems can reduce the environmental impact of waste and support the city's sustainability goals.
2. **Resource Efficiency**: The circular economy can enhance resource efficiency in Raipur's industrial sector by encouraging the use of recycled materials, reducing waste generation, and adopting sustainable production practices. This approach can help alleviate resource scarcity and support the long-term sustainability of industrial activities.
3. **Sustainable Urbanization**: Integrating circular economy principles into urban planning can support sustainable urbanization by promoting green infrastructure, preserving natural habitats, and improving waste management in rapidly growing urban areas. Circular practices can help mitigate the environmental impact of urban expansion and improve the quality of life for residents.
4. **Economic Resilience**: Adopting circular economy practices can drive economic resilience by fostering innovation, creating new business opportunities, and supporting the development of sustainable industries. Circular economy models can enhance the city's economic diversification and contribute to long-term prosperity.

In conclusion, the socio-economic context of Raipur, characterized by its industrial profile and urbanization trends, underscores the need for a circular economy approach. By addressing waste management challenges, resource scarcity, and environmental degradation through circular principles, Raipur can achieve sustainable development and enhance the city's resilience and quality of life.

1.5 Implementation of Circular Economy Principles in Raipur

Raipur, as a growing industrial and urban center, faces significant environmental and resource management challenges. Implementing circular economy principles can help address these challenges by designing out waste and pollution, keeping products and materials in use, and regenerating natural systems. This section will focus on strategies for designing out waste and pollution in Raipur, specifically promoting eco-friendly building materials, reducing single-use plastics, and adopting cleaner production technologies.

1.5.1 Strategies for Designing Out Waste and Pollution

Designing out waste and pollution involves rethinking how products and processes are designed to minimize environmental impact from the outset. The construction industry in Raipur, like in many other cities, is a major consumer of raw materials and a significant contributor to waste and pollution. Transitioning to eco-friendly building materials is essential for minimizing the environmental impact of construction activities.

1. **Use of Recycled and Renewable Materials**: Promoting the use of recycled materials, such as recycled concrete and steel, can significantly reduce the demand for virgin raw materials and the associated environmental impacts. Additionally, renewable materials like bamboo, which has a high growth rate and carbon sequestration ability, can be used as an alternative to traditional building materials (Kibert, 2016).
2. **Sustainable Design Practices**: Encouraging sustainable design practices, such as modular construction and prefabrication, can minimize waste generation during construction. These methods allow for the precise use of materials, reducing off-cuts and excess waste. Prefabrication also enables the reuse of components, aligning with the principles of the circular economy (Ghaffar et al., 2020).
3. **Incentives for Green Building Certifications**: Providing incentives for builders and developers to achieve green building certifications, such as LEED (Leadership in Energy and Environmental Design) or IGBC (Indian Green Building Council) certification, can drive the adoption of eco-friendly building

practices. These certifications encourage the use of sustainable materials and construction techniques, thereby reducing waste and pollution (IGBC, 2022).

4. **Local Sourcing of Materials**: Promoting the use of locally sourced materials can reduce the carbon footprint associated with transportation and support local economies. Using indigenous materials that are adapted to the local climate and environment can also enhance the sustainability of buildings (Hussain & Oosthuizen, 2013).

1.5.2 Reducing Single-Use Plastics

Single-use plastics are a significant source of waste and pollution in urban environments. Strategies to reduce their use in Raipur can involve a combination of regulatory measures, public awareness campaigns, and the promotion of alternatives.

1. **Regulatory Measures**: Implementing strict regulations to ban or limit the use of single-use plastics is a crucial step. For example, the Chhattisgarh government has already banned plastic bags, but enforcement and extension of such bans to other single-use plastic items can further reduce plastic waste (Government of Chhattisgarh, 2018).
2. **Promoting Alternatives**: Encouraging the use of alternatives to single-use plastics, such as reusable bags, biodegradable packaging, and metal or glass containers, can significantly decrease plastic waste. Public campaigns and partnerships with local businesses to provide incentives for using these alternatives can enhance their adoption (UNEP, 2018a, b).
3. **Public Awareness Campaigns**: Educating the public about the environmental impact of single-use plastics and the benefits of using alternatives is essential. Campaigns can be conducted through various media channels, community programs, and schools to raise awareness and encourage behavioral change (WWF, 2020).
4. **Extended Producer Responsibility (EPR)**: Implementing EPR policies that hold manufacturers accountable for the lifecycle of their plastic products can incentivize the design of more sustainable products and packaging. Companies can be required to take back used products and ensure their proper recycling or disposal (OECD, 2021a, b).

1.5.3 Cleaner Production Technologies

Cleaner production technologies are essential for reducing waste and pollution at the source in industrial processes. These technologies focus on improving efficiency, reducing emissions, and minimizing resource consumption.

1. **Energy Efficiency Improvements**: Encouraging industries to adopt energy-efficient technologies and practices can significantly reduce emissions and

energy consumption. For instance, upgrading to energy-efficient machinery, optimizing production processes, and utilizing renewable energy sources can lead to substantial environmental benefits (IEA, 2021).

2. **Water Conservation and Recycling**: Implementing water conservation measures and recycling technologies can reduce water consumption and minimize wastewater generation in industrial processes. Techniques such as closed-loop water systems, water-efficient machinery, and wastewater treatment and reuse can help industries in Raipur manage water resources more sustainably (UNIDO, 2020).
3. **Waste Minimization and Resource Recovery**: Promoting waste minimization practices and resource recovery technologies can reduce the environmental impact of industrial waste. Industries can adopt practices such as lean manufacturing, which focuses on reducing waste and improving efficiency, and implement technologies that recover valuable materials from waste streams (EPA, 2019).
4. **Pollution Control Technologies**: Encouraging the adoption of advanced pollution control technologies, such as air scrubbers, electrostatic precipitators, and biofilters, can help industries reduce their emissions and comply with environmental regulations. These technologies can capture and neutralize pollutants before they are released into the environment (Kumar & Bhardwaj, 2021).
5. **Green Chemistry and Sustainable Materials**: Promoting the use of green chemistry principles and sustainable materials in industrial processes can reduce the generation of hazardous waste and the environmental impact of chemical manufacturing. Green chemistry focuses on designing chemical products and processes that minimize the use and generation of hazardous substances (Anastas & Warner, 1998).

1.5.4 Case Studies and Examples

1. **Eco-Friendly Construction Projects**: The implementation of eco-friendly building materials can be illustrated by projects such as the IGBC-certified green buildings in Raipur. These buildings use recycled materials, energy-efficient designs, and sustainable construction practices to minimize their environmental footprint (IGBC, 2022).
2. **Plastic Ban Success Stories**: The success of plastic bans in other Indian cities, such as Mumbai and Delhi, can serve as a model for Raipur. These cities have implemented comprehensive bans on single-use plastics, coupled with strong enforcement and public awareness campaigns, leading to significant reductions in plastic waste (CPCB, 2019).
3. **Cleaner Production in Industries**: The adoption of cleaner production technologies by industries in Raipur can be demonstrated through examples such as the implementation of energy-efficient furnaces and waste heat recovery systems

in steel plants. These technologies have helped reduce energy consumption and emissions while improving production efficiency (IEA, 2021).

1.5.5 Policy Recommendations

To support the implementation of circular economy principles in Raipur, the following policy recommendations are proposed:

1. **Strengthening Regulations and Enforcement**: Implementing and enforcing regulations that promote the use of eco-friendly materials, reduce single-use plastics, and mandate cleaner production technologies is crucial. This includes providing clear guidelines, monitoring compliance, and imposing penalties for non-compliance (Government of Chhattisgarh, 2018).
2. **Providing Financial Incentives**: Offering financial incentives, such as tax breaks, grants, and subsidies, for businesses and industries that adopt circular economy practices can encourage the transition to more sustainable operations. Incentives can be targeted at investments in eco-friendly materials, cleaner technologies, and waste reduction initiatives (OECD, 2021a, b).
3. **Promoting Research and Development**: Supporting research and development in circular economy technologies and practices can drive innovation and improve the feasibility of sustainable solutions. Funding for R&D initiatives, collaboration with academic institutions, and establishing innovation hubs can foster the development of new technologies (UNIDO, 2020).
4. **Enhancing Public-Private Partnerships**: Encouraging collaboration between the public and private sectors can facilitate the implementation of circular economy principles. Public-private partnerships can support infrastructure development, resource recovery projects, and the promotion of sustainable business practices (WWF, 2020).
5. **Engaging the Community**: Involving the community in circular economy initiatives through awareness campaigns, educational programs, and participatory projects can build public support and encourage sustainable behaviors. Community engagement can also help identify local challenges and develop tailored solutions (UNEP, 2018a, b).

Implementing circular economy principles in Raipur is essential for addressing the city's environmental and resource management challenges. By promoting eco-friendly building materials, reducing single-use plastics, and adopting cleaner production technologies, Raipur can design out waste and pollution, enhance resource efficiency, and improve environmental sustainability. The strategies outlined in this section, supported by robust policies and community engagement, can drive the transition towards a circular economy in Raipur, contributing to the city's long-term resilience and prosperity.

1.6 Initiatives to Keep Products and Materials in Use

Keeping products and materials in use is a core principle of the circular economy, aimed at extending the lifecycle of resources through reuse, repair, refurbishment, and recycling. Implementing this principle in Raipur involves various community and business-driven initiatives.

1.6.1 Community Recycling Programs

Community recycling programs play a crucial role in ensuring that materials are kept in use and not discarded as waste. Effective recycling programs can significantly reduce the amount of waste that ends up in landfills and can provide raw materials for new products.

1. **Establishing Recycling Centers**: Setting up well-organized recycling centers in various neighborhoods can make it convenient for residents to recycle household waste. These centers can accept a range of materials, including paper, plastics, glass, and metals, and ensure they are properly sorted and sent to recycling facilities (Gupta & Arora, 2020).
2. **Educational Campaigns**: Public awareness campaigns about the importance of recycling and how to do it correctly can increase participation rates. Schools, community groups, and local media can be engaged to disseminate information and encourage responsible waste disposal practices (UNEP, 2018a, b).
3. **Incentive Programs**: Introducing incentive programs, such as deposit-return schemes for bottles and cans or reward points for recycling certain materials, can motivate residents to participate actively in recycling efforts. These programs can be supported by local government and private sector partnerships (WRAP, 2019).
4. **Integration with Informal Sector**: In many Indian cities, the informal sector plays a significant role in recycling. Integrating informal recyclers into formal recycling programs can enhance the efficiency and reach of these initiatives. Providing training, protective gear, and fair compensation can help professionalize the sector and improve recycling rates (Chaturvedi et al., 2020a, b).

1.6.2 Repair Cafes and Second-Hand Markets

Repair cafes and second-hand markets are community-driven initiatives that encourage the repair and reuse of products, thereby extending their lifecycle and reducing waste.

1. **Establishing Repair Cafes**: Repair cafes are pop-up events where volunteers with repair skills help residents fix their broken items, ranging from electronics

to clothing and furniture. These cafes not only reduce waste but also foster community spirit and skill-sharing (Bakx, 2020).

2. **Promoting Second-Hand Markets**: Second-hand markets provide a platform for buying and selling used goods, which can significantly reduce the demand for new products and the resources required to produce them. Regularly organized flea markets, online platforms, and thrift stores can help create a thriving second-hand economy (Williams, 2020).
3. **Skill Development Workshops**: Offering workshops on repair skills, such as electronics repair, sewing, and furniture restoration, can empower residents to fix their own items and reduce dependence on new products. These workshops can be held at community centers, schools, and repair cafes (Koch & Hamelin, 2021).
4. **Partnerships with Manufacturers**: Collaborating with manufacturers to provide spare parts, repair manuals, and training can support the repair and reuse of products. Manufacturers can also be encouraged to design products that are easier to repair and upgrade (Stahel, 2016).

1.7 Efforts to Regenerate Natural Systems

Regenerating natural systems is essential for restoring the ecological balance and enhancing the resilience of urban environments. In Raipur, efforts to regenerate natural systems can include enhancing green spaces, promoting urban agriculture, and implementing reforestation and soil conservation projects.

1.7.1 Enhancing Green Spaces

Green spaces, such as parks, gardens, and green belts, are vital for the ecological and social health of urban areas. Enhancing green spaces in Raipur can provide multiple benefits, including improved air quality, biodiversity, and recreational opportunities.

1. **Developing Urban Parks**: Establishing and maintaining urban parks in densely populated areas can provide residents with accessible green spaces for recreation and relaxation. These parks can be designed with native plant species to support local biodiversity (Chiesura, 2004).
2. **Green Roofs and Walls**: Encouraging the installation of green roofs and walls on buildings can increase the amount of greenery in the city, reduce urban heat island effects, and improve air quality. Incentives and guidelines for green building practices can support this initiative (Berardi et al., 2014).
3. **Community Gardens**: Establishing community gardens where residents can grow their own fruits, vegetables, and flowers can enhance urban green spaces and promote social interaction. Community gardens can also serve as educational sites for sustainable agriculture practices (Armstrong, 2000).

4. **Tree Planting Programs**: Large-scale tree planting programs can increase urban tree cover, sequester carbon, and provide shade and cooling effects. These programs can be organized in collaboration with local NGOs, schools, and community groups (Nowak, 2018).

1.7.2 Promoting Urban Agriculture

Urban agriculture involves growing food within city limits, utilizing various spaces such as rooftops, vacant lots, and community gardens. Promoting urban agriculture in Raipur can enhance food security, reduce food miles, and provide green spaces.

1. **Rooftop Gardens**: Encouraging the use of rooftops for gardening can turn underutilized spaces into productive areas for growing vegetables, herbs, and fruits. Rooftop gardens can also provide insulation for buildings and reduce energy consumption (Specht et al., 2014).
2. **Vertical Farming**: Implementing vertical farming techniques, which involve growing crops in stacked layers, can maximize the use of limited urban space and increase food production. Vertical farms can be established in urban centers and integrated with existing buildings (Despommier, 2011).
3. **Hydroponics and Aquaponics**: Promoting hydroponic and aquaponic systems, which use nutrient-rich water instead of soil to grow plants, can allow for efficient urban agriculture with minimal space requirements. These systems can be set up in urban areas to produce fresh, local food (Junge et al., 2017).
4. **Support for Local Farmers**: Providing support to local farmers through subsidies, access to markets, and training in sustainable practices can enhance urban and peri-urban agriculture. Farmers' markets and community-supported agriculture (CSA) programs can connect urban residents with local producers (Feenstra, 1997).

1.7.3 Reforestation and Soil Conservation Projects

Reforestation and soil conservation projects are critical for restoring degraded ecosystems, enhancing biodiversity, and improving soil health. In Raipur, these efforts can mitigate the effects of urbanization and support sustainable land management.

1. **Reforestation Initiatives**: Implementing reforestation projects in and around Raipur can restore deforested areas, increase tree cover, and sequester carbon. These initiatives can involve planting native tree species that are well-adapted to the local climate and ecosystem (Chazdon, 2008).
2. **Soil Conservation Techniques**: Promoting soil conservation techniques, such as contour plowing, terracing, and the use of cover crops, can prevent soil ero-

sion and improve soil fertility. These techniques can be implemented in agricultural areas and urban gardens to enhance soil health (Lal, 2015).

3. **Watershed Management**: Effective watershed management practices can improve water quality, reduce runoff, and support reforestation efforts. Activities such as constructing check dams, restoring wetlands, and protecting riparian zones can enhance the resilience of local water systems.
4. **Community Involvement**: Engaging the community in reforestation and soil conservation projects through volunteer programs, educational workshops, and awareness campaigns can foster a sense of stewardship and collective action. Community participation can enhance the success and sustainability of these initiatives (Pretty, 2003).

Implementing circular economy principles in Raipur requires a multifaceted approach that includes keeping products and materials in use through community recycling programs and repair cafes, as well as regenerating natural systems by enhancing green spaces, promoting urban agriculture, and implementing reforestation and soil conservation projects. These initiatives, supported by robust policies, community engagement, and innovative practices, can drive Raipur towards a sustainable and resilient future.

1.8 Case Studies of Circular Economy Initiatives in Raipur

1.8.1 Raipur Municipal Corporation's Waste Management Program

The Raipur Municipal Corporation (RMC) has been at the forefront of managing urban waste and promoting sustainable practices in the city. Given the rapid urbanization and industrial growth in Raipur, effective waste management is critical to minimizing environmental impact and enhancing the quality of life for its residents.

1. **Integrated Waste Management System**: RMC has implemented an integrated waste management system that includes the segregation of waste at the source, efficient collection, transportation, and disposal. This system ensures that recyclable materials are separated from organic waste, reducing the burden on landfills and facilitating recycling processes (Raipur Municipal Corporation, 2020).
2. **Door-to-Door Collection**: The corporation has established a door-to-door waste collection service that covers both residential and commercial areas. This service is complemented by awareness campaigns that educate citizens on proper waste segregation and disposal practices. The collected waste is then transported to processing facilities for recycling or composting (Raipur Municipal Corporation, 2021).

3. **Composting and Biomethanation Plants**: To manage organic waste, RMC has set up composting and biomethanation plants. These facilities convert organic waste into compost and biogas, respectively. The compost is used as a soil conditioner in urban gardens and agricultural fields, while biogas is utilized as a renewable energy source (Government of Chhattisgarh, 2020).
4. **Material Recovery Facilities (MRFs)**: RMC operates several MRFs where recyclable materials are sorted, cleaned, and processed for sale to recycling industries. These facilities play a crucial role in diverting recyclable waste from landfills and promoting a circular economy (Raipur Municipal Corporation, 2021).
5. **Public Participation and Education**: Engaging the public in waste management efforts is essential for the success of the program. RMC conducts regular workshops, school programs, and community meetings to raise awareness about the importance of waste segregation, recycling, and responsible consumption (Raipur Municipal Corporation, 2020).

1.8.2 Green Buildings and Sustainable Construction Practices

Green buildings and sustainable construction practices are essential for reducing the environmental footprint of the construction industry. In Raipur, several initiatives have been undertaken to promote green building practices, with notable examples such as the Chhattisgarh State Renewable Energy Development Agency (CREDA) office building.

1.8.3 Chhattisgarh State Renewable Energy Development Agency (CREDA) Office Building

The CREDA office building in Raipur is a prime example of sustainable construction and green building practices. This building incorporates several eco-friendly features that minimize its environmental impact and enhance energy efficiency.

1. **Energy-Efficient Design**: The CREDA office building is designed to maximize natural light and ventilation, reducing the need for artificial lighting and air conditioning. The building's orientation and window placement ensure optimal use of daylight, while energy-efficient lighting and HVAC systems further reduce energy consumption (CREDA, 2021).
2. **Renewable Energy Integration**: The building is equipped with solar panels that generate a significant portion of its electricity needs. This integration of renewable energy sources not only reduces the building's carbon footprint but also serves as a demonstration of CREDA's commitment to promoting renewable energy in the state (Chhattisgarh Renewable Energy Development Agency, 2020).
3. **Water Conservation Measures**: To address water scarcity, the CREDA office building incorporates rainwater harvesting systems and low-flow fixtures. The

harvested rainwater is used for landscaping and other non-potable purposes, reducing the demand for municipal water supply (CREDA, 2021).

4. **Sustainable Materials**: The construction of the CREDA office building utilized sustainable materials, including recycled and locally sourced materials. This approach minimizes the environmental impact associated with the extraction and transportation of building materials (Chhattisgarh Renewable Energy Development Agency, 2020).
5. **Green Certification**: The building has received green building certification, recognizing its adherence to sustainable construction practices and energy efficiency standards. This certification serves as a benchmark for other construction projects in Raipur to follow (CREDA, 2021).

CREDA Office Building

1.8.4 Recycling and Upcycling Initiatives

Recycling and upcycling initiatives are crucial components of the circular economy, helping to keep products and materials in use and reducing waste. In Raipur, startups and community-driven initiatives are making significant strides in this area, with Greenovation being a notable example.

1.8.5 Greenovation Startup

Greenovation is a Raipur-based startup that focuses on recycling and upcycling waste materials into innovative and sustainable products. The startup aims to address waste management challenges while promoting a culture of sustainability and creativity.

1. **Product Innovation**: Greenovation specializes in creating products from recycled materials, such as furniture, home decor, and accessories. These products are designed to be aesthetically pleasing and functional, demonstrating that waste materials can be transformed into valuable resources (Greenovation, 2021).
2. **Community Engagement**: The startup engages with the local community through workshops and events that promote recycling and upcycling. These activities educate residents on the importance of waste reduction and encourage them to participate in sustainable practices (Greenovation, 2021).
3. **Partnerships and Collaborations**: Greenovation collaborates with local businesses, schools, and NGOs to collect recyclable materials and spread awareness about the benefits of upcycling. These partnerships help to expand the startup's reach and impact, fostering a community-wide commitment to sustainability (Greenovation, 2021).
4. **Economic Empowerment**: By providing training and employment opportunities in the field of recycling and upcycling, Greenovation contributes to the economic empowerment of local residents. The startup's initiatives help to create jobs and support the development of green skills in the community (Greenovation, 2021).
5. **Circular Economy Model**: Greenovation operates on a circular economy model, where waste materials are continuously repurposed and reused. This approach not only reduces waste but also conserves resources and minimizes environmental impact (Greenovation, 2021).

The implementation of circular economy principles in Raipur is evident through the efforts of the Raipur Municipal Corporation's waste management program, green building practices exemplified by the CREDA office building, and innovative recycling and upcycling initiatives like Greenovation. These initiatives collectively contribute to reducing waste, conserving resources, and promoting sustainability in the city. By continuing to support and expand these efforts, Raipur can set a benchmark for other urban centers to follow in their journey towards a circular economy.

Devpuri Tarnsfer Station, Raipur

1.9 Benefits of Adopting the Circular Economy in Raipur

Implementing circular economy principles in Raipur can yield significant environmental, economic, and social benefits. These advantages extend beyond immediate impacts, fostering long-term sustainability and resilience.

1.9.1 Environmental Benefits

The environmental benefits of a circular economy are vast, including waste reduction, resource conservation, and lower carbon emissions. These benefits are crucial for addressing the ecological challenges posed by urbanization and industrial activities in Raipur.

1.9.2 Waste Reduction

One of the most immediate and visible benefits of a circular economy is the reduction of waste. By emphasizing reuse, recycling, and upcycling, the amount of waste sent to landfills is significantly decreased.

1. **Extended Product Lifespan**: Initiatives such as repair cafes and second-hand markets help extend the lifespan of products, reducing the frequency with which items are discarded. This not only lessens the waste burden but also reduces the need for new products (Bocken et al., 2016).
2. **Efficient Waste Management**: Enhanced waste management systems, including community recycling programs and material recovery facilities (MRFs), ensure that recyclable materials are effectively separated and processed. This minimizes the environmental impact of waste disposal (Ghisellini et al., 2016).
3. **Reduction in Landfill Use**: By diverting waste from landfills through composting and recycling initiatives, the amount of waste that ends up in these sites is reduced. This alleviates the pressure on existing landfill spaces and mitigates associated environmental problems, such as soil and water contamination (Geng et al., 2019).

1.9.3 Resource Conservation

A circular economy promotes the efficient use of resources, ensuring that materials are used to their fullest potential and minimizing the need for virgin resource extraction.

1. **Recycling and Upcycling**: By transforming waste materials into new products, recycling and upcycling initiatives reduce the demand for raw materials. This

conserves natural resources and reduces the environmental impact of extraction and processing (Kirchherr et al., 2017).

2. **Sustainable Construction Practices**: The use of eco-friendly building materials and sustainable construction methods, as seen in the CREDA office building, conserves resources and reduces waste. Green buildings also incorporate renewable energy and water conservation measures, further preserving natural resources (Kibert, 2016).
3. **Urban Agriculture**: Promoting urban agriculture through rooftop gardens and vertical farming helps utilize urban spaces for food production. This reduces the need for agricultural land expansion and conserves resources such as water and soil (Specht et al., 2014).

1.9.4 Lower Carbon Emissions

Reducing carbon emissions is a critical aspect of mitigating climate change. The circular economy contributes to lower emissions through various means.

1. **Reduced Production Needs**: By extending the lifespan of products and promoting recycling, the need for new production is decreased, leading to lower energy consumption and fewer emissions associated with manufacturing processes (Stahel, 2016).
2. **Renewable Energy Integration**: The incorporation of renewable energy sources, such as solar panels on the CREDA office building, reduces reliance on fossil fuels and lowers greenhouse gas emissions. Renewable energy projects can be scaled up to cover larger areas of the city (Chhattisgarh Renewable Energy Development Agency, 2020).
3. **Sustainable Transportation**: Encouraging the use of bicycles, public transport, and electric vehicles, as part of a broader circular economy strategy, can reduce emissions from the transportation sector. Improved waste management and resource efficiency also contribute to lower emissions (Geissdoerfer et al., 2017).

1.9.5 Economic Benefits

Implementing circular economy principles can lead to substantial economic benefits, including cost savings, job creation, and new business opportunities.

1.9.6 Cost Savings

A circular economy can reduce costs for businesses and consumers by optimizing resource use and minimizing waste.

1. **Reduced Material Costs**: Businesses that adopt circular practices, such as recycling and reusing materials, can save on raw material costs. This is particularly relevant for industries in Raipur where raw material prices can be volatile (Ellen MacArthur Foundation, 2013).
2. **Lower Waste Management Costs**: Efficient waste management systems reduce the costs associated with waste collection, transportation, and disposal. By diverting waste from landfills, cities can save on landfill maintenance and expansion expenses (Ritzén & Sandström, 2017).
3. **Energy Savings**: Energy-efficient buildings and renewable energy integration result in lower utility bills for businesses and residents. The CREDA office building, with its solar panels and energy-saving features, exemplifies how green buildings can lead to significant cost savings (Chhattisgarh Renewable Energy Development Agency, 2020).

1.9.7 Job Creation

The transition to a circular economy can create new jobs in various sectors, contributing to economic growth and development.

1. **Recycling and Waste Management**: The expansion of recycling programs and waste management facilities creates jobs in sorting, processing, and logistics. Formalizing the informal recycling sector can also provide stable employment and improve working conditions (Chaturvedi et al., 2020a, b).
2. **Green Building and Construction**: The demand for sustainable construction practices generates employment opportunities in the construction industry, including roles in design, engineering, and building. Training programs in green building techniques can support this job creation (Kibert, 2016).
3. **Repair and Maintenance Services**: Initiatives like repair cafes and second-hand markets stimulate demand for skilled labor in repair and refurbishment services. This can create jobs for technicians, craftsmen, and entrepreneurs (Bakx, 2020).
4. **Agriculture and Urban Farming**: Promoting urban agriculture can generate jobs in farming, marketing, and distribution. Community-supported agriculture (CSA) programs and farmers' markets can provide additional employment opportunities (Feenstra, 1997).

1.9.8 New Business Opportunities

The circular economy opens up new business opportunities, fostering innovation and entrepreneurship.

1. **Circular Startups**: Startups focused on recycling, upcycling, and sustainable products, such as Greenovation, can thrive in a circular economy. These busi-

nesses can develop innovative products and services that cater to environmentally conscious consumers (Greenovation, 2021).
2. **Green Technology and Services**: The demand for green technologies and services, including renewable energy systems, waste management solutions, and sustainable construction materials, creates opportunities for new businesses and investments (OECD, 2021a, b).
3. **Sharing Economy Platforms**: Platforms that facilitate the sharing and reuse of goods, such as peer-to-peer rental services and second-hand marketplaces, can flourish in a circular economy. These platforms can reduce the need for new products and promote resource efficiency (Williams, 2020).

1.9.9 Social Benefits

The circular economy also offers numerous social benefits, enhancing consumer value, fostering community development, and improving overall quality of life.

1.9.10 Enhanced Consumer Value

Consumers benefit from higher quality, more durable products and services in a circular economy.

1. **Cost-Effective Products**: Second-hand markets and refurbished products offer consumers access to high-quality goods at lower prices. This increases affordability and reduces the financial burden on households (Williams, 2020).
2. **Extended Product Warranties**: Products designed for longevity and repairability often come with extended warranties, providing consumers with greater value and peace of mind. This encourages a shift away from disposable culture (Stahel, 2016).
3. **Access to Repair Services**: The availability of repair services through initiatives like repair cafes ensures that consumers can maintain and extend the life of their products. This reduces waste and promotes a culture of care and sustainability (Bakx, 2020).

1.9.11 Community Development

Community-driven initiatives in the circular economy can strengthen social ties and enhance local development.

1. **Community Recycling Programs**: Engaging residents in recycling programs fosters a sense of community responsibility and collective action. These pro-

grams can also improve neighborhood cleanliness and aesthetics (Gupta & Arora, 2020).

2. **Urban Agriculture Projects**: Community gardens and urban farming initiatives provide spaces for social interaction, education, and collaboration. These projects can strengthen community bonds and promote healthy lifestyles (Armstrong, 2000).
3. **Local Economic Development**: By supporting local businesses and creating jobs, circular economy initiatives contribute to the economic development of communities. This can lead to improved infrastructure, services, and overall quality of life (Pretty, 2003).

The implementation of circular economy principles in Raipur brings substantial environmental, economic, and social benefits. Waste reduction, resource conservation, and lower carbon emissions address critical environmental challenges. Economic advantages, such as cost savings, job creation, and new business opportunities, drive sustainable development. Social benefits, including enhanced consumer value and community development, improve the quality of life for residents. Together, these benefits demonstrate the transformative potential of the circular economy in fostering a sustainable and resilient future for Raipur.

1.10 Challenges to Implementing the Circular Economy

1.10.1 Awareness and Education

Raising awareness and providing education on the circular economy are fundamental steps in promoting sustainable practices in Raipur. Public awareness campaigns using mass media—such as television, radio, newspapers, and social media—can effectively disseminate information about the importance of waste reduction, recycling, and sustainable consumption. Community workshops and seminars offer hands-on training and practical knowledge, targeting various demographics including students, homemakers, and local businesses, fostering a deeper understanding of circular economy principles. Integrating these concepts into school and university curricula ensures that the younger generation grows up with a strong foundation in sustainability, preparing them to be proactive participants in the circular economy. Additionally, partnerships with NGOs and environmental organizations can amplify educational efforts, creating a more informed and engaged community (Ellen MacArthur Foundation, 2015; UNEP, 2018a, b).

1.10.2 Policy and Regulatory Support

Effective policy and regulatory support are crucial for the successful implementation of a circular economy in Raipur. Governments at local, state, and national levels need to establish clear guidelines and regulations that encourage and enforce sustainable practices. Policies that incentivize recycling, waste reduction, and the use of eco-friendly materials can drive businesses and individuals towards more sustainable behavior. Regulatory frameworks should also include penalties for non-compliance with waste management and environmental standards to ensure accountability. Government subsidies and tax incentives for green technologies and sustainable practices can further encourage businesses to adopt circular economy models. In addition, setting targets for waste reduction, recycling rates, and renewable energy usage can provide clear goals for the city to achieve (OECD, 2021a, b).

1.10.3 Infrastructure and Technology

Developing the necessary infrastructure and advancing technological capabilities are essential for supporting the circular economy in Raipur. Investment in modern waste management facilities, including recycling centers, composting units, and material recovery facilities, is vital. These facilities should be equipped with advanced technologies to efficiently process and manage waste. Implementing smart waste management systems, such as IoT-enabled waste bins and real-time monitoring, can optimize waste collection and reduce operational costs. Additionally, promoting the use of renewable energy sources and energy-efficient technologies in industrial processes can significantly reduce the environmental impact. The construction of green buildings, as demonstrated by the CREDA office building, showcases the integration of sustainable materials and renewable energy systems, serving as a model for future projects (Chhattisgarh Renewable Energy Development Agency, 2020).

1.10.4 Collaboration and Partnerships

Collaboration and partnerships among various stakeholders are key to fostering a circular economy in Raipur. Public-private partnerships can leverage the strengths and resources of both sectors to develop and implement sustainable initiatives. Collaboration with local businesses, industries, and startups can drive innovation and create new market opportunities for circular products and services. Engaging NGOs, community groups, and environmental organizations can enhance grassroots participation and ensure that the benefits of the circular economy reach all segments of society. International cooperation and knowledge exchange with cities and organizations that have successfully implemented circular economy practices can provide valuable insights and best practices. Creating platforms for dialogue and collaboration among stakeholders can facilitate the sharing of ideas, resources, and expertise, thereby accelerating the transition to a circular economy (Geng et al., 2019).

By focusing on awareness and education, policy and regulatory support, infrastructure and technology, and collaboration and partnerships, Raipur can effectively implement circular economy principles and move towards a more sustainable and resilient future.

1.11 Policy Recommendations for Promoting the Circular Economy in Raipur

To foster the transition to a circular economy in Raipur, targeted policy recommendations are crucial. These recommendations should focus on providing incentives for sustainable practices, establishing standards for waste management and recycling, implementing green procurement policies, and promoting public awareness through campaigns and educational programs. By drawing on successful examples such as the CREDA office building, these policies can guide Raipur toward a more sustainable future.

1.11.1 Incentives for Sustainable Practices

1.11.1.1 Financial Incentives and Tax Benefits

To encourage businesses and individuals to adopt sustainable practices, the government should offer financial incentives such as tax breaks, subsidies, and grants. For instance, providing tax rebates for companies that implement recycling programs or use eco-friendly materials can stimulate investment in circular economy initiatives. Grants can be allocated for research and development of innovative technologies that promote resource efficiency and waste reduction. Inspired by successful models like the CREDA office building, which benefits from state renewable energy subsidies, Raipur can implement similar financial incentives to support sustainable construction and renewable energy projects (Chhattisgarh Renewable Energy Development Agency, 2020).

1.11.1.2 Support for Green Startups

Creating an enabling environment for green startups is essential for fostering innovation. This can be achieved through funding opportunities, incubators, and accelerators specifically focused on circular economy ventures. By providing seed funding and mentorship to startups working on recycling technologies, waste reduction solutions, and sustainable products, Raipur can stimulate entrepreneurship in the green sector.

1.11.2 Standards for Waste Management and Recycling

1.11.2.1 Establishing Comprehensive Waste Management Standards

The city should develop and enforce comprehensive waste management standards that include guidelines for waste separation, collection, and disposal. These standards should mandate the segregation of recyclable materials at the source, ensuring that waste is sorted effectively before reaching processing facilities. This approach improves the efficiency of recycling processes and reduces contamination, leading to higher quality recyclables (Geng et al., 2019).

1.11.2.2 Implementation of Extended Producer Responsibility (EPR)

Extended Producer Responsibility (EPR) programs can be implemented to make manufacturers accountable for the entire lifecycle of their products, including post-consumer waste. This policy requires producers to take responsibility for the collection, recycling, and disposal of their products. By enforcing EPR, Raipur can ensure that companies contribute to waste management efforts and reduce the environmental impact of their products.

1.11.2.3 Setting Recycling Targets and Metrics

Establishing clear recycling targets and performance metrics is vital for tracking progress and ensuring accountability. The city should set ambitious yet achievable recycling targets and regularly monitor progress through data collection and reporting. These targets should be aligned with national and international standards to promote consistency and effectiveness in waste management (OECD, 2021a, b).

1.11.3 Green Procurement Policies

1.11.3.1 Adoption of Green Procurement Guidelines

The city government and public institutions should adopt green procurement policies that prioritize the purchase of environmentally friendly products and services. This includes selecting products made from recycled materials, ensuring that suppliers adhere to sustainability standards, and choosing services that minimize environmental impact. By setting an example through green procurement, public institutions can drive demand for sustainable products and encourage businesses to follow suit.

1.11.3.2 Promotion of Sustainable Construction Practices

Green procurement policies should extend to construction projects, ensuring that sustainable building materials and technologies are used. The CREDA office building serves as a model for incorporating renewable energy systems and eco-friendly materials. Similar policies can be implemented for all new construction and renovation projects in Raipur, requiring adherence to green building standards and certification programs.

1.11.4 Public Awareness Campaigns and Educational Programs

1.11.4.1 Comprehensive Public Awareness Campaigns

Public awareness campaigns are essential for informing residents and businesses about the benefits and practices of the circular economy. Campaigns should utilize various media channels, including social media, radio, and community events, to reach a broad audience. Key messages should focus on the importance of waste reduction, recycling, and sustainable consumption, with actionable tips and information on local recycling programs and services.

1.11.4.2 Educational Programs in Schools and Communities

Integrating circular economy concepts into school curricula and community education programs is crucial for building long-term awareness and engagement. Schools should incorporate lessons on sustainability and the circular economy into science and social studies subjects, while community organizations can offer workshops and training sessions on recycling, composting, and sustainable living. Partnering with local businesses and environmental groups can enhance the effectiveness of these educational programs (UNEP, 2018a, b).

1.11.4.3 Collaboration with NGOs and Environmental Groups

Collaborating with non-governmental organizations (NGOs) and environmental groups can amplify public awareness efforts and provide additional resources and expertise. These organizations can assist in organizing community events, educational workshops, and outreach programs, leveraging their experience and networks to engage a wider audience and promote circular economy principles.

By implementing these policy recommendations—offering incentives for sustainable practices, establishing standards for waste management and recycling, adopting green procurement policies, and promoting public awareness and education—Raipur can effectively advance its transition to a circular economy. Drawing on the successful example of the CREDA office building, these strategies will support the development of a more sustainable and resilient urban environment, paving the way for a greener future.

1.12 Conclusion

The transition to a circular economy represents a critical shift towards sustainable development, particularly for rapidly growing cities like Raipur. As Raipur confronts significant economic and environmental challenges—such as waste management issues, resource scarcity, and environmental degradation—embracing the principles of the circular economy offers a pathway to address these problems comprehensively. By focusing on strategies that include incentivizing sustainable practices, establishing robust waste management standards, implementing green procurement policies, and fostering public awareness and education, Raipur can lay the foundation for a more resilient and sustainable urban environment.

Incentivizing sustainable practices through financial rewards and support for green startups can stimulate widespread adoption of circular economy principles. These incentives will not only drive innovation but also help integrate circular practices into the mainstream business model, creating a more sustainable economic framework. Standards for waste management and recycling, including the implementation of Extended Producer Responsibility (EPR) programs, will enhance the efficiency of waste processing and encourage companies to take responsibility for their products throughout their lifecycle.

Green procurement policies will further embed sustainability into public projects and construction practices, setting an example for private enterprises and residents. The successful integration of these policies, as evidenced by initiatives such as the CREDA office building, highlights the potential for transformative change when sustainability is prioritized in procurement and construction.

Public awareness and educational programs are equally crucial in ensuring the long-term success of the circular economy. By raising awareness and providing practical education on sustainability, Raipur can foster a culture of environmental stewardship among its citizens. These programs will empower individuals and businesses to make informed decisions, support recycling initiatives, and adopt sustainable consumption practices.

Collaboration and partnerships with NGOs, environmental groups, and the private sector will amplify these efforts, creating a collaborative approach to sustainability that leverages diverse expertise and resources. Such partnerships can enhance community engagement and drive collective action towards achieving circular economy goals.

Lastly, the effective implementation of these strategies will not only address Raipur's current economic and environmental challenges but also pave the way for a more sustainable and resilient future. By embracing the circular economy, Raipur has the opportunity to become a leading example of urban sustainability, demonstrating how cities can thrive economically while preserving and regenerating their natural environment. The journey towards a circular economy is complex, but with committed leadership, strategic policies, and active community involvement, Raipur can successfully navigate this transition and reap the numerous environmental, economic, and social benefits that it offers.

References

Anastas, P. T., & Warner, J. C. (1998). *Green chemistry: Theory and practice*. Oxford University Press.

Armstrong, D. (2000). A survey of community gardens in upstate New York: Implications for health promotion and community development. *Health & Place, 6*(4), 319–327.

Bakx, K. (2020). Repair cafes: A fix for throwaway culture? *CBC News*. Retrieved from https://www.cbc.ca/news/canada/calgary/repair-cafes-throwaway-culture-1.5433216

Berardi, U., GhaffarianHoseini, A., & GhaffarianHoseini, A. (2014). State-of-the-art analysis of the environmental benefits of green roofs. *Applied Energy, 115*, 411–428.

Bocken, N. M. P., de Pauw, I., Bakker, C., & van der Grinten, B. (2016). Product design and business model strategies for a circular economy. *Journal of Industrial and Production Engineering, 33*(5), 308–320.

Brundtland Commission. (1987). *Our common future*. Oxford University Press.

Census of India. (2011). *Primary census abstract*. Retrieved from http://censusindia.gov.in

Central Pollution Control Board (CPCB). (2021). *Report on Municipal solid waste management in India*. Retrieved from https://cpcb.nic.in/municipal-solid-waste-management/

Chaturvedi, B., Arora, R., & Saluja, M. S. (2020a). Integrating the informal recycling sector in solid waste management in India. *Waste Management & Research, 38*(9), 983–990.

Chaturvedi, B., Arora, R., & Saluja, M. S. (2020b). Recycling in India: Elimination of informality for sustainable development. *Economic & Political Weekly, 55*(35), 55–63.

Chazdon, R. L. (2008). Beyond deforestation: Restoring forests and ecosystem services on degraded lands. *Science, 320*(5882), 1458–1460.

Chhattisgarh Renewable Energy Development Agency. (2020). *CREDA annual report 2019–2020*. Retrieved from https://creda.cgstate.gov.in

Chiesura, A. (2004). The role of urban parks for the sustainable city. *Landscape and Urban Planning, 68*(1), 129–138.

CPCB. (2019). *Annual report for the year 2018–19 on implementation of plastic waste management rules, 2016*. Central Pollution Control Board. Retrieved from https://cpcb.nic.in

CREDA. (2021). *Green building initiatives*. Retrieved from https://creda.cgstate.gov.in/green-building

Despommier, D. (2011). The vertical farm: Controlled environment agriculture carried out in tall buildings would create a sustainable urban food supply. *Journal of Consumer Protection and Food Safety, 6*, 233–236.

Ellen MacArthur Foundation. (2013). *Towards the circular economy: Economic and business rationale for an accelerated transition*. Retrieved from https://www.ellenmacarthurfoundation.org/publications/towards-the-circular-economy-vol-1-an-economic-and-business-rationale-for-an-accelerated-transition

Ellen MacArthur Foundation. (2015). *Delivering the circular economy: A toolkit for policymakers*. Retrieved from https://www.ellenmacarthurfoundation.org/publications/delivering-the-circular-economy-a-toolkit-for-policymakers

EPA. (2019). *Lean manufacturing and the environment: Research on advanced manufacturing systems and the environment and recommendations for leveraging better environmental performance*. United States Environmental Protection Agency. Retrieved from https://www.epa.gov

Feenstra, G. (1997). Local food systems and sustainable communities. *American Journal of Alternative Agriculture, 12*(1), 28–36.

Geissdoerfer, M., Savaget, P., Bocken, N. M. P., & Hultink, E. J. (2017). The circular economy—A new sustainability paradigm? *Journal of Cleaner Production, 143*, 757–768.

Geng, Y., Sarkis, J., Ulgiati, S., & Zhang, P. (2019). Measuring China's circular economy. *Science, 347*(6227), 1076–1077.

Ghaffar, S. H., Corker, J., & Fan, M. (2020). Additive manufacturing technology and its implementation in construction as an eco-innovative solution. *Automation in Construction, 93*, 1–11.

Ghisellini, P., Cialani, C., & Ulgiati, S. (2016). A review on circular economy: The expected transition to a balanced interplay of environmental and economic systems. *Journal of Cleaner Production, 114*, 11–32.

Government of Chhattisgarh. (2018). *Notification on ban of plastic bags in Chhattisgarh*. Retrieved from https://www.cgstate.gov.in

Government of Chhattisgarh. (2020). *Waste management strategies*. Retrieved from https://www.cgstate.gov.in

Greenovation. (2021). *About us*. Retrieved from https://greenovation.com

Gupta, N., & Arora, R. (2020). Public participation in solid waste management: A case study of Delhi. *Environmental Management and Sustainable Development, 9*(1), 93–108.

Hussain, M., & Oosthuizen, P. (2013). Sustainable construction: Role of green building rating systems. In *Proceedings of the International conference on sustainable design and construction*.

IEA. (2021). *Energy Efficiency 2021*. International Energy Agency. Retrieved from https://www.iea.org

IGBC. (2022). *Green building rating systems*. Indian Green Building Council. Retrieved from https://igbc.in

Junge, R., König, B., Villarroel Walker, R., & Komives, T. (2017). Strategic points in aquaponics. *Water, 9*(3), 182.

Kibert, C. J. (2016). *Sustainable construction: Green building design and delivery*. Wiley.

Kirchherr, J., Reike, D., & Hekkert, M. (2017). Conceptualizing the circular economy: An analysis of 114 definitions. *Resources, Conservation and Recycling, 127*, 221–232.

Koch, D., & Hamelin, L. (2021). Circular economy strategies for electric vehicle batteries reduce reliance on raw materials. *Nature Communications, 12*, 4582.

Kumar, S., & Bhardwaj, R. (2021). Advanced air pollution control technologies: Current status and challenges. *Environmental Pollution, 273*, 116441.

Lal, R. (2015). Restoring soil quality to mitigate soil degradation. *Sustainability, 7*(5), 5875–5895.

Nowak, D. J. (2018). Understanding the benefits and costs of urban forest ecosystems. In S. T. A. Pickett, M. L. Cadenasso, & B. McGrath (Eds.), *Resilience in ecology and urban design* (pp. 137–156). Springer.

OECD. (2021a). *Extended producer responsibility: Updated guidance for efficient waste management*. Organisation for Economic Co-operation and Development. Retrieved from https://www.oecd.org

OECD. (2021b). *The circular economy in cities and regions*. Retrieved from https://www.oecd.org

Pretty, J. (2003). Social capital and the collective management of resources. *Science, 302*(5652), 1912–1914.

Raipur Municipal Corporation. (2020). *Integrated waste management system*. Retrieved from https://www.rmc.cg.gov.in

Raipur Municipal Corporation. (2021). *Annual report 2020–2021*. Retrieved from https://www.rmc.cg.gov.in

Ritzén, S., & Sandström, G. Ö. (2017). Barriers to the circular economy—Integration of perspectives and domains. *Procedia CIRP, 64*, 7–12.

Specht, K., Siebert, R., Hartmann, I., Freisinger, U. B., Sawicka, M., Werner, A., Thomaier, S., Henckel, D., Walk, H., & Dierich, A. (2014). Urban agriculture of the future: An overview of sustainability aspects of food production in and on buildings. *Agriculture and Human Values, 31*(1), 33–51.

Stahel, W. R. (2016). The circular economy. *Nature News, 531*(7595), 435–438.

UNEP. (2018a). *Single-use plastics: A roadmap for sustainability*. United Nations Environment Programme. Retrieved from https://www.unep.org

UNEP. (2018b). *The role of education and awareness in the transition towards a circular economy*. Retrieved from https://www.unep.org/resources/report/role-education-and-awareness-transition-towards-circular-economy

UNIDO. (2020). *Circular economy: An introduction and its application to industrial wastewater management*. United Nations Industrial Development Organization. Retrieved from https://www.unido.org

Williams, C. (2020). The rise of second-hand fashion: How vintage, resale, and thrift stores are booming. *The Guardian*. Retrieved from https://www.theguardian.com

WRAP. (2019). *Incentives to stimulate recycling: A report for the Welsh Government*. Waste & Resources Action Programme. Retrieved from https://www.wrap.org.uk

WWF. (2020). *Plastic pollution: Causes and consequences*. World Wildlife Fund. Retrieved from https://www.worldwildlife.org

Chapter 2
Transforming Trash to Treasure: Pioneering Work of the Women's Groups of Shillong (India) Towards Achieving the Circular Economy

Soma Mitra, Shrabantee Sar, B. P. Sahu, and Srikanta Patnaik

Abstract In the Harmonious hills of Shillong, Meghalaya women's group has revolutionized waste management by transforming garbage into saleable compost. By collecting and composting organic waste from households and markets, they produce high-quality, eco-friendly fertilizer. This initiative began with a keen awareness of the environmental issues posed by the city's growing waste problem. Recognizing the potential to turn this problem into an opportunity, the group started by collecting organic waste from households and local markets. Their work represents the principles of a circular economy, turning waste into a valuable resource for local farmers and gardeners. The diverse initiatives undertaken by women's groups in Shillong, ranging from community clean-up drives to educational workshops on waste management. It explores the economic, social, and environmental impacts of these initiatives, demonstrating how they have led to job creation, improved community health, and significant reductions in pollution. Beside the remarkable progress, the article also addresses the challenges these groups face, such as financial constraints and infrastructural limitations. It discusses the innovative strategies employed to overcome these obstacles, including community engagement and partnerships with local governments and NGOs. In conclusion, the article underscores the importance of continued support for these grassroots movements and highlights the potential for scaling up successful initiatives to broader regions, ultimately contributing to a more sustainable and equitable future.

S. Mitra (✉) · S. Sar
Department of Humanities & Social Sciences, Siksha 'O' Anusandhan (Deemed to be University), Bhubaneswar, India

B. P. Sahu
Adult and Continuing Education, North Eastern Hill University, Shillong, India

S. Patnaik
IIMT, Bhubaneswar, India

S. Patnaik et al. (eds.), *Circular Economy and Sustainable Value Creation through Eco-Innovation*, Circular Economy and Sustainability,
https://doi.org/10.1007/978-3-031-88516-7_2

Keywords Circular economy · Waste management · Sustainable development · Recycling · Upcycling · Environmental conservation · Organic farming · Composting · Socio-economic Impact

2.1 Introduction

Dedicated women's group in Shillong has revolutionized (Banerjee, 2010) waste management by transforming garbage into saleable compost. By collecting and composting organic waste from households and markets, they produce high-quality, eco-friendly fertilizer. This initiative not only reduces landfill waste but also provides a sustainable income for the women involved. The collected waste is meticulously sorted (Banerjee, 2010), with organic material being set aside for composting. Using traditional composting techniques alongside modern innovations, the women ensure that the organic waste decomposes efficiently (Banerjee, 2010). This process is carefully monitored, with regular turning and moisture checks, to produce high-quality compost. Once the (Banerjee, 2010) compost is ready, it is packaged and sold to local farmers and gardeners. This not only provides an eco-friendly alternative to chemical fertilizers but also creates a (Banerjee, 2010) sustainable revenue stream for the women involved. The group's efforts have significantly reduced the amount of waste sent to landfills, lowering the environmental impact and promoting soil health through the use of natural fertilizers (Baruah, 2011). The initiative has empowered the women economically and socially, giving them a sense of purpose and pride in their work. Their success has inspired other communities to adopt similar practices, highlighting the feasibility and benefits of a circular economy model (Baruah, 2011). By converting waste into a valuable resource, the women's group in Shillong has shown that sustainable practices can lead to both environmental and economic gains, paving the way for a greener and more resilient future.

2.2 Understanding the Circular Economy

The (Bhowmik, 2013) circular economy is an innovative economic model designed to minimize waste and make the most of available resources. Unlike the traditional linear economy, which follows a "take, make, dispose" pattern, the (Bhowmik, 2013) circular economy aims to close the loop by rethinking production and consumption processes. The core principle is to design products and systems that enable the continuous reuse, recycling, and regeneration of materials. At its core, the circular economy focuses on creating a system where resources are kept in use for as long as possible. It involves (Bhowmik, 2013) designing products for durability, repairability, and recyclability, thus reducing the need for raw materials and energy. By doing so, it seeks to decouple economic growth from resource consumption and environmental impact (Bhowmik, 2013). In practice, the circular economy encourages businesses and consumers to adopt practices such as product sharing, leasing, and re-manufacturing. For example, rather than purchasing a new item, consumers might opt to lease a product or buy a refurbished version. This approach not only extends the (Bhowmik, 2013) lifespan of products but also reduces the volume of waste generated. Moreover, the circular economy promotes the development of closed-loop supply chains where waste from one process becomes input for another. The transition to a circular economy also has significant economic benefits. It can drive innovation, create new business models, and open up opportunities for economic growth. Additionally, it supports the creation of green jobs and provides a more sustainable approach to resource management. Moreover, the circular economy represents a shift towards a more sustainable, regenerative system that aligns economic activities with (Basu, 2012) environmental management. By redefining how we produce, consume, and manage resources, it offers a pathway to reduce waste, conserve natural resources, and build a more adaptable economy for the future.

2.2.1 Importance of Waste Management and Recycling

Banerjee, 2010 Effective waste management and recycling are crucial for sustaining environmental health and promoting resource conservation (Banerjee, 2010). Proper waste management helps reduce pollution, mitigate the impact of waste on natural ecosystems, and protect human health by minimizing exposure to hazardous materials (Banerjee, 2010). Recycling plays a key role in conserving natural resources by transforming waste products into valuable materials that can be reused in manufacturing processes (Basu, 2012). It only reduces the demand for raw materials but also lowers energy consumption and greenhouse gas emissions associated with production (Bhowmik, 2013). Furthermore, durable waste management and (Baruah, 2011) recycling programs support economic growth by creating jobs and providing innovation in recycling technologies and sustainable practices. Prioritizing waste management and recycling is essential for building a cleaner, more sustainable future for our planet.

2.2.2 The Role of Women's Groups in Sustainable Practices

(Basu, 2012) Women's groups play an important role in promoting sustainable practices globally, driving community-based initiatives that address environmental and social challenges. These groups are often at the forefront of efforts to conserve natural resources, manage waste, and promote recycling, utilizing their unique insights and leadership within their communities (Bhowmik, 2013). By implementing sustainable agricultural practices, advocating for clean energy, and fostering eco-friendly business ventures, women's groups contribute significantly to environmental stewardship and economic resilience (Basu, 2012). Their involvement in education and advocacy also raises awareness about the importance of sustainability, inspiring broader community participation (Basu, 2012). The empowerment and active participation of women's groups are essential in achieving sustainable development goals and creating a more equitable and environmentally conscious world.

2.3 Women's Self-Help Groups (SHGs) in Shillong

(Choudhury, 2014) Women's Self-Help Groups (SHGs) in Shillong represent a transformative force in community development and empowerment. These groups, composed primarily of local women, have become instrumental in providing economic independence and social change. Through SHGs, women gain access to microfinance, enabling them to start or expand small businesses. It economically not only support and enhances their financial stability but also contributes to household incomes and local economic growth. In addition to economic benefits, SHGs play a vital role in community development. They organize health camps, literacy programs, and educational workshops, addressing essential needs and improving the overall well-being of the community (Choudhury, 2014). These initiatives promote health awareness, educational advancement, and social cohesion. Environmental sustainability is another area where SHGs make a significant impact. Many groups engage in activities such as tree planting, waste management, and organic farming (Bhowmik, 2013). These efforts help preserve local ecosystems and promote eco-friendly practices within the community (Choudhury, 2014). Despite facing challenges such as limited resources and societal resistance, the achievements of SHGs in Shillong are remarkable (Bhowmik, 2013). Their contributions to economic empowerment, community development, and environmental conservation underscore their role in shaping a more equitable and sustainable future for the region.

2.3.1 Initiatives Undertaken by Women's Self-Help Groups (SHGs) in Shillong

(Das, 2015) In Shillong, women's groups have undertaken on several commendable initiatives to enhance community development and economic empowerment. These groups, often organized as self-help collectives, focus on skill development, sustainable agriculture, and small-scale entrepreneurship (Choudhury, 2014). They conduct workshops on weaving, handicrafts, and organic farming, which not only preserve traditional knowledge but also provide women with sustainable income sources. Additionally, they engage in environmental conservation efforts, such as tree planting and waste management, promoting ecological awareness within the community (Choudhury, 2014). Health and education are also key areas of focus, with women's groups organizing health camps and literacy programs to improve the overall well-being of families. Through microfinance schemes, they offer financial support to start or expand small businesses, providing economic independence (Choudhury, 2014). These initiatives collectively contribute to the socio-economic upliftment of women in Shillong, enhancing their roles as active participants in the community's growth and development.

2.3.2 *Impact of Women's Self-Help Groups (SHG) in Shillong*

(Deka, 2016) Women's Self-Help Groups (SHGs) in Shillong have emerged as powerful agents for socio-economic change and environmental sustainability. These grassroots organizations, composed predominantly of women from various socio-economic backgrounds, have made significant developments in improving the local community and environment through their diverse initiatives (Ghosh, 2018). Their impact can be observed in areas such as economic empowerment, community development, and environmental conservation (Das, 2015). By organizing themselves into self-help groups, women gain access to microfinance, which enables them to start or expand small businesses. This financial support helps them break free from the cycle of poverty and enhances their economic independence. The groups provide training in various skills such as weaving, handicrafts, and organic farming, which not only preserve traditional knowledge but also open new avenues for income generation. These activities have led to increased household incomes and improved living standards, as women can contribute more significantly to their families' financial stability. These initiatives address critical needs within the community, such as health awareness and educational opportunities (Das, 2015). Health camps conducted by SHGs offer essential medical services and information, improving public health and well-being. Literacy programs focus on enhancing educational levels, which is essential for long-term socio-economic development (Das, 2015). Moreover, SHGs facilitate networking and solidarity among women, strengthening social cohesion and empowering them to address common issues collectively.

2.3.3 *Challenges Faced and the Strategies Employed*

2.3.3.1 Challenges Faced

(Ghosh, 2018) Women's Self-Help Groups (SHGs) in Shillong face several significant challenges that hinder their effectiveness. Limited financial resources and inadequate access to capital often restrict their ability to expand and sustain their initiatives (Deka, 2016). Infrastructural deficiencies, such as poor transportation and lack of suitable facilities, further constrain their operations. Moreover, societal resistance deep-rooted in traditional norms and gender biases can undermine the influence and participation of women in these groups. Skill gaps and insufficient awareness about SHG benefits also pose obstacles to their growth and success.

(i) ***Limited Resources:*** (Deka, 2016) Many Self-Help Groups (SHGs) in Shillong struggle with inadequate financial resources and lack of access to capital, which hampers their ability to scale up their initiatives or undertake new projects.
(ii) ***Infrastructural Deficiencies:*** (Deka, 2016) Insufficient infrastructure, such as poor transportation and inadequate facilities for meetings and training, limits the effectiveness and reach of SHGs.
(iii) ***Societal Resistance:*** (Deka, 2016) Traditional societal norms and gender biases sometimes undermine the authority and influence of women, creating barriers to their active participation and acceptance in various community activities.
(iv) ***Skill Gaps:*** (Deka, 2016) There are often gaps in the technical and managerial skills required to run businesses or manage projects effectively, affecting the overall success of SHG initiatives.
(v) ***Lack of Awareness:*** (Ghosh, 2018) Limited awareness about the benefits and operations of SHGs among potential members can impede the growth and effectiveness of these groups.

2.3.3.2 Strategies Employed

(Gogoi, 2019) Women's Self-Help Groups (SHGs) in Shillong utilize microfinance schemes and partner with financial institutions to secure necessary funding (Hazarika, 2020). They engage in capacity-building programs and organize skill development workshops to enhance members' entrepreneurial and managerial skills. Forming alliances with NGOs, government agencies, and others which helps them to access additional resources and support services. Furthermore, they conduct community outreach and awareness campaigns to change societal attitudes and promote gender equality, increasing acceptance and support for their activities.

(i) ***Microfinance and Capacity Building:*** (Gogoi, 2019) To address resource constraints, SHGs leverage microfinance schemes and collaborate with financial institutions to secure funding. They also engage in capacity-building programs to enhance members' skills in business management and financial literacy.
(ii) ***Partnerships and Networking:*** (Ghosh, 2018) SHGs form alliances with non-governmental organizations (NGOs), government agencies, and other stakeholders to access additional resources, training, and support services.
(iii) ***Community Engagement:*** (Gogoi, 2019) By conducting awareness campaigns and community outreach, SHGs work to change societal attitudes and promote gender equality, thereby increasing acceptance and support for their activities.

(iv) ***Skill Development Workshops:*** (Ghosh, 2018) To bridge skill gaps, SHGs organize workshops and training sessions focused on entrepreneurial skills, leadership, and technical knowledge relevant to their projects.
(v) ***Infrastructure Improvement:*** (Gogoi, 2019) SHGs advocate for better infrastructure and utilize available resources to improve their facilities and access to essential services, thereby enhancing their operational effectiveness.

2.4 Circular Economy by Women's Groups in Shillong

(Gogoi, 2019) Women's groups in Shillong have been at the forefront of promoting a circular economy, focusing on sustainable practices through the principles of reducing, reusing, recycling, recovering, and regenerating (Ghosh, 2018). Their initiatives not only contribute to environmental conservation but also empower women economically and socially.

2.4.1 Reduce: Minimizing Waste at the Source

(Hazarika, 2020) One of the primary goals of the circular economy is to minimize waste production (Gogoi, 2019). The **Lum Sohpetbneng Women's Group** has spearheaded a campaign to reduce plastic usage in their community (Gogoi, 2019). By promoting the use of cloth bags, encouraging bulk buying to minimize packaging waste, and organizing workshops on sustainable consumption, they have significantly reduced the amount of plastic waste generated. This initiative not only helps the environment but also educates the community on the importance of mindful consumption.

2.4.2 Reuse: Extending the Life of Products

Reusing materials is another critical aspect of a circular economy (Mawlynnong Women's Cooperative, 2018). The **Mawlynnong Women's Cooperative** has established a community saving store where residents can donate and purchase gently used clothing, household items, and tools. This initiative reduces the demand for new products and extends the lifecycle of existing ones. Moreover, the proceeds from the saving store are reused into community projects, providing a sense of communal responsibility and cooperation.

2.4.3 Recycle: Turning Waste into Resources

(Hazarika, 2020) Recycling is essential for converting waste into reusable materials. The Shillong Waste Warriors, a group of women dedicated to waste management, have implemented a comprehensive recycling program in their locality (Gogoi, 2019). They collect, sort, and recycle various types of waste, including paper, glass, and metal. Their efforts have not only reduced landfill waste but also created job opportunities for women in the recycling industry. The group's success has inspired other communities to adopt similar practices.

2.4.4 Recover: Harnessing Value from Waste

(Hazarika, 2020) Recovering value from waste is another pillar of the circular economy (Hazarika, 2020). The **Laitkynsew Women's Association** has introduced biogas plants that convert organic waste into energy (Hazarika, 2020). By collecting food scraps and agricultural waste, they produce biogas used for cooking and heating. This initiative not only provides a renewable energy source but also reduces greenhouse gas emissions from decomposing organic waste. The biogas project has improved the community's energy security and reduced reliance on non-renewable energy sources.

2.4.5 Regenerate: Restoring and Renewing Resources

(Hazarika, 2020) Regeneration involves restoring and renewing natural resources (Hazarika, 2020). The **Sohra Women's Environmental Group** has taken significant steps in reforestation and soil regeneration. They have planted thousands of native trees in degraded areas and implemented soil conservation techniques such as terracing and composting. These efforts have led to the restoration of local ecosystems, improved biodiversity, and enhanced soil health, ensuring long-term sustainability for future generations.

2.5 Notable Contributions by Women's Group in Circular Economy: Case Studies

2.5.1 Lum Sohpetbneng Women's Group

(Kharmawphlang, 2021) The Lum Sohpetbneng Women's Group in Shillong, Meghalaya founded in 2012 was initiated by a small group of local women driven by the desire to address environmental challenges and create sustainable livelihoods

(Hazarika, 2020). Their mission is deeply rooted in the Khasi philosophy of living in harmony with nature, which emphasizes respect for natural resources, community welfare, and intergenerational equity. The group aims to reduce waste, promote resource efficiency, and enhance the socio-economic status of women through various eco-innovative practices. The group stands as a pioneer of eco-innovation and the circular economy in northeastern India. The group is named after the sacred peak Lum Sohpetbneng, which holds immense cultural and spiritual significance for the indigenous Khasi people, embodies the principles of sustainability, community resilience, and women's empowerment (Kharmawphlang, 2021). The Lum Sohpetbneng Women's Group has significantly impacted the local community, particularly in terms of women's empowerment. By providing training, resources, and support, the group has enabled many women to become financially independent and contribute to their households and communities. The skills and knowledge gained through the group's initiatives have also boosted the confidence and self-esteem of the women involved (Hazarika, 2020). Its approach to the circular economy involves several key practices: *reduce, reuse, recycle, recover, and regenerate*. By focusing on waste minimization and resource recovery, they have developed a comprehensive model that transforms waste into valuable resources. The group represents the power of grassroots movements in driving environmental sustainability and social change (Hazarika, 2020). Their dedication to eco-innovation and the circular economy not only addresses pressing environmental issues but also empowers women, strengthens community bonds, and fosters a culture of sustainability for future generations.

2.5.1.1 Waste Reduction and Reuse

(Kharmawphlang, 2021) One of their primary initiatives is the reduction of single-use plastics and other non-biodegradable materials (Kharmawphlang, 2021). The group conducts awareness campaigns, workshops, and community outreach programs to educate the local population about the environmental impact of plastic waste. They encourage the use of reusable bags, containers, and other eco-friendly alternatives.

2.5.1.2 Recycling Initiatives

Kharmawphlang, 2021 The Lum Sohpetbneng Women's Group has established a robust recycling program that involves collecting, sorting, and processing various types of waste. They have set up collection centers where community members can drop off recyclable materials such as paper, plastic, glass, and metal (Kharmawphlang, 2021). These materials are then sold to recycling companies, generating income for the group and supporting their other initiatives.

2.5.1.3 Resource Recovery and Upcycling

(Kharmawphlang, 2021) In addition to recycling, the group engages in upcycling, transforming waste materials into new, useful products. They create handmade crafts, jewelry, and home decor items from discarded materials, which are sold at local markets and online platforms. This not only reduces waste but also provides a source of income for the women involved.

2.5.1.4 Organic Farming and Composting

(Kharmawphlang, 2021) The group's commitment to sustainability extends to organic farming and composting. They promote the use of organic fertilizers and pesticides, encouraging local farmers to adopt eco-friendly agricultural practices. The women also run community composting programs, turning kitchen waste into nutrient-rich compost that is used to improve soil fertility and increase crop yields.

2.5.2 Mawlynnong Women's Cooperative

(Mawlynnong Women's Cooperative, 2018) The Mawlynnong Women's Cooperative was formed in 2005, inspired by the community's long-standing traditions of cleanliness and harmony with nature. Nestled in the enchanting East Khasi Hills of Meghalaya, the Mawlynnong Women's Cooperative stands as a testament to the power of collective action and sustainable living (Mawlynnong Women's Cooperative, 2018). Mawlynnong, often hailed as the "cleanest village in Asia," owes much of its acclaim to the relentless efforts of this cooperative, which has become a model of eco-consciousness and women's empowerment (Mawlynnong Women's Cooperative, 2018). The cooperative's founding members were driven by a vision to enhance the well-being of their village while preserving its natural beauty (Mawlynnong Women's Cooperative, 2018). Their mission encompasses environmental sustainability, economic development, and social empowerment, with a particular focus on uplifting women. It has profoundly impacted the local community, particularly in empowering women. By involving them in decision-making processes and providing opportunities for skill development and income generation, the cooperative has enhanced the social and economic status of women in the village. This empowerment has immense effects, improving the overall quality of life and providing a sense of community pride and solidarity (Mawlynnong Women's Cooperative, 2018). The Group's commitment to environmental preservation, economic empowerment, and cultural heritage has made Mawlynnong a shining example of sustainable living. Through their efforts, the cooperative not only upholds the village's reputation as the cleanest in Asia but also creates a resilient, self-reliant community that can inspire others far beyond the borders of Meghalaya.

2.5.2.1 Sustainability Initiatives

(Kharmawphlang, 2021) Central to the cooperative's efforts are various sustainability initiatives that align with the principles of the circular economy: reducing waste, reusing resources, recycling materials, and promoting regenerative practices.

2.5.2.2 Waste Management and Cleanliness

(Mawlynnong Women's Cooperative, 2018) One of the cooperative's flagship initiatives is its rigorous waste management program. Members organize regular clean-up drives, ensuring that every household practice waste segregation. Biodegradable waste is composted, while non-biodegradable waste is collected for recycling (Mawlynnong Women's Cooperative, 2018). This meticulous approach has kept Mawlynnong free of litter, setting a high standard for cleanliness and environmental stewardship.

2.5.2.3 Eco-Friendly Tourism

(Mawlynnong Women's Cooperative, 2018) The cooperative has channeled Mawlynnong's reputation for cleanliness to promote eco-friendly tourism. By developing homestays and nature trails, they offer visitors an immersive experience of the village's pristine environment and cultural heritage (Mawlynnong Women's Cooperative, 2018). The cooperative trains local women to manage these tourism services, providing them with a steady income while preserving the village's ecological integrity.

2.5.2.4 Organic Farming and Herbal Medicine

(Mishra, 2011) Agriculture is another cornerstone of the cooperative's activities. The women practice organic farming, cultivating a variety of crops without chemical fertilizers or pesticides. They also grow medicinal herbs, which are used to produce traditional remedies. These practices not only ensure food security and health benefits but also conserve the region's rich biodiversity.

2.5.2.5 Handicrafts and Upcycling

(Mishra, 2011) The cooperative promotes the creation of handicrafts from natural and recycled materials. Women craft items such as bamboo baskets, woven textiles, and decorative pieces, which are sold to tourists and through online platforms. This initiative reduces waste and provides an additional source of income, helping to sustain the cooperative's activities.

2.5.3 *Shillong Waste Warriors*

(Mishra, 2011) Shillong Waste Warriors was established in 2014 by a group of environmentally conscious individuals who recognized the urgent need to address waste management issues in the city. Their vision is to create a waste-free Shillong by inculcating a culture of cleanliness, sustainability, and environmental stewardship among its residents Mishra, 2011. They aim to achieve this through a combination of community engagement, education, and innovative waste management solutions. Shillong, the capital of Meghalaya, known for its scenic beauty and pleasant climate, has faced increasing challenges related to waste management due to rapid urbanization and population growth (Nongkynrih, 2012). Beside these challenges, the Shillong Waste Warriors have emerged as a powerful force dedicated to handling waste issues and promoting sustainable practices. The dynamic group is not just cleaning up the city but also educating and empowering the community to embrace a cleaner, greener future. It had made a significant social impact by empowering local communities to take charge of their environment Nongkynrih, 2012. Through their initiatives, they have created a sense of ownership and pride among residents regarding their city's cleanliness and sustainability. The group's efforts have also led to increased civic participation and volunteerism, as more people join the cause and contribute to the collective goal of a cleaner Shillong. The Waste Warriors' approach to waste management is comprehensive and multidimensional, involving various initiatives that address different aspects of the waste problem.

2.5.3.1 Community Clean-Up Drives

(Mishra, 2011) One of the most visible and impactful activities of the Shillong Waste Warriors is their regular community clean-up drives. These events, often organized in collaboration with local schools, colleges, businesses, and residents, involve cleaning public spaces such as parks, streets, and markets (Mishra, 2011). The clean-up drives not only remove litter but also raise awareness about the importance of proper waste disposal and the impact of littering on the environment.

2.5.3.2 Waste Segregation and Recycling

(Nongkynrih, 2012) Promoting waste separation at the source is a key focus for the Waste Warriors. They conduct workshops and awareness campaigns to educate households and businesses about the importance of separating biodegradable and non-biodegradable waste (Mishra, 2011). The group has set up waste separation units in various neighbourhood and collaborates with recycling companies to ensure that recyclable materials are processed appropriately. This initiative reduces the burden on landfills and encourages recycling and reuse.

Fig. 2.1 (Nongkynrih, 2012) Members of the Iainehskhem Self-Help Group (SHG) as Waste Warriors

2.5.3.3 Educational Programs

(Nongkynrih, 2012) *Education* is at the heart of the Shillong Waste Warriors' mission. They conduct environmental education programs in schools and colleges, teaching students about waste management, recycling, and the broader impacts of waste on the environment (Mishra, 2011). These programs are designed to inculcate a sense of responsibility and environmental consciousness in the younger generation, ensuring that future leaders are well-equipped to continue the fight against waste.

2.5.3.4 Innovative Waste Solutions

(Nongkynrih, 2012) The Waste Warriors are also known for their innovative approaches to waste management. They have introduced composting programs to turn organic waste into valuable compost for gardening and agriculture. Additionally, they promote the use of eco-friendly alternatives to single-use plastics, such as cloth bags and biodegradable packaging materials (Nongkynrih, 2012). These innovations help reduce the overall waste generated and encourage sustainable consumption habits. A live example by the Members of the Iainehskhem Self-Help Group (SHG) in Fig. 2.1.

2.5.4 Laitkynsew Women's Association

(Qabeer, 2018) The Laitkynsew Women's Association was established in 2010 by a group of enterprising women who recognized the need for collective action to address environmental and socio-economic challenges. Nestled in the lush green hills of Meghalaya, the Laitkynsew Women's Association has become a light of hope and progress for its community (Nongkynrih, 2012). This association, based in the village of Laitkynsew near Shillong, represents the trans-formative power of grassroots initiatives led by women. Their work is a demonstration to the potential for eco-innovation, sustainable development, and social empowerment to coexist and progress. The founding members were driven by a vision to uplift their community by inculcating the power of unity and sustainable practices. The association's initiatives have had a profound impact on the local community. Women have gained financial independence, improved their living standards, and become more active participants in community affairs. The emphasis on sustainable practices has also led to a cleaner, healthier environment, benefiting all residents of Laitkynsew (Nongkynrih, 2012). Their commitment to Eco-innovation, sustainability, and women's empowerment has not only improved the quality of life in Laitkynsew but also serves as an inspiration for other communities. Through their efforts, the association is paving the way for a brighter, more sustainable future, demonstrating the power of collective action and the invincible spirit of women. The association's mission includes environmental conservation, economic development, and the empowerment of women through various innovative and community-driven projects.

Eco-Innovation and Sustainability Initiatives: (Qabeer, 2018) Central to the association's efforts several Eco-innovation and sustainability initiatives reflects the principles of the circular economy: *reducing waste, reusing resources, recycling materials, and promoting regenerative practices.*

2.5.4.1 Waste Management and Recycling

(Chhakchhuak, 2024) One of the association's flagship initiatives is its comprehensive waste management program. They have implemented a system of waste segregation at the source, educating households on the importance of separating biodegradable and non-biodegradable waste (Qabeer, 2018). Biodegradable waste is composted and used to enhance soil fertility in local farms, while non-biodegradable waste is collected for recycling. This initiative not only keeps the village clean but also promotes a culture of recycling and waste reduction.

2.5.4.2 Organic Farming and Agroforestry

(Qabeer, 2018) Agriculture plays a significant role in the lives of the people in Laitkynsew, and the Women's Association has introduced organic farming and agroforestry practices to promote sustainability. By avoiding chemical fertilizers and pesticides, they ensure healthier crops and a safer environment (Chhakchhuak, 2024). They also engage in agroforestry, integrating trees and shrubs into agricultural landscapes, which helps in soil conservation, enhances biodiversity, and provides additional sources of income through the sale of timber and non-timber forest products.

2.5.4.3 (Qabeer, 2018) Handicrafts and Upcycling

The women of the association are skilled artisans who create beautiful handicrafts from locally sourced and recycled materials. These products, which include baskets, textiles, and decorative items, are sold at local markets and online platforms (Qabeer, 2018). The upcycling of waste materials into valuable products not only reduces environmental impact but also provides a significant source of income for the women involved, promoting economic self-sufficiency.

2.5.4.4 Community Education and Empowerment

Education and (Qabeer, 2018) empowerment is at the core of the Laitkynsew Women's Association's mission. They conduct regular workshops and training sessions on various topics such as sustainable farming, waste management, financial literacy, and leadership skills (Chhakchhuak, 2024). These programs are designed to equip women with the knowledge and skills needed to improve their livelihoods and contribute to community development. By (Chhakchhuak, 2024) providing a sense of empowerment, the association helps women take active roles in decision-making processes within their households and the community. Ambitious, large-scale plans remain unfulfilled, but a women's self-help group composed of waste pickers turned recyclers could provide a viable solution to the hill station's severe waste crisis. A live example is illustrated in Fig. 2.2.

2.6 Impact on the Community

2.6.1 Economic Impact

(Qabeer, 2018) Women's groups in Shillong have been instrumental in fostering economic empowerment through various entrepreneurial ventures. These groups often engage in activities such as handicrafts, organic farming, and small-scale manufacturing, creating a wave that enhances both individual and community wealth (Qabeer, 2018). By utilizing local resources and traditional skills, they have managed to establish sustainable businesses that generate income, provide employment, and improve the overall economic stability of the region (Qabeer, 2018). A key aspect of the economic impact of women's groups in Shillong is their commitment to circular economy principles. Unlike the traditional linear economy, which follows a ***'take-make-dispose' model***, the circular economy emphasizes the

Fig. 2.2 (Chhakchhuak, 2024) Women's self-help group comprised of waste pickers-turned-recyclers could help the hill station solve its ugly waste crisis

importance of resource efficiency, waste reduction, and recycling. Women's groups have embraced these principles, pioneering initiatives that contribute to both economic growth and environmental sustainability.

(Qabeer, 2018) Moreover, many women's groups are involved in recycling and upcycling projects. They collect and process waste materials, transforming them into valuable products. This approach not only reduces the volume of waste that ends up in landfills but also creates new economic opportunities. By turning waste into marketable goods, these groups demonstrate how circular economy practices can drive economic innovation while addressing environmental challenges. Additionally, some women's groups focus on organic farming and sustainable agriculture (Chhakchhuak, 2024). By employing eco-friendly farming techniques and promoting the use of organic fertilizers, they reduce the reliance on harmful chemicals and contribute to healthier ecosystems. The produce from these farms is often sold in local markets or used in community-supported agriculture programs, thereby supporting both environmental sustainability and local economies (Qabeer, 2018). By pioneering circular economy practices, these groups not only contribute to local economic development but also promote environmental stewardship. Their success stories serve as powerful examples of how women's leadership and community engagement can drive meaningful change.

One notable example is the success of local handicraft enterprises. Women's groups, performing in the traditional weaving and textile skills, have turned these crafts into lucrative businesses. They produce and market a range of products, from traditional shawls to contemporary apparel, attracting both domestic and international buyers. This not only provides a steady source of income for the women involved but also stimulates local tourism and supports the broader economy. The craftsmanship is exquisitely detailed and visually stunning, earning acclaim both in India and internationally. In Meghalaya, crafting items from cane and bamboo has become a primary livelihood for many tribal communities. A live example is illustrated in Fig. 2.3.

Fig. 2.3 (Qabeer, 2018) Handicraft works by women group of Shillong

2.6.2 *Social Impact*

(Chhakchhuak, 2024) The social impact of women's groups in Shillong is also evident in their role in providing community cohesion and resilience. By working on projects that address local needs—*such as waste management, sustainable agriculture, and community-based recycling programs*—these groups contribute to a stronger and more interconnected community. Their activities often involve collaboration with local residents, schools, and other organizations, creating a network of support and shared purpose (Chhakchhuak, 2024). Women's groups in Shillong are at the forefront of promoting the circular economy, utilizing their collective efforts to create positive social change. One of the most significant impacts of these groups is the empowerment of women. By engaging in circular economy practices such as recycling, upcycling, and sustainable agriculture, women are gaining financial independence and social status. This empowerment extends beyond economic benefits, providing a sense of self-worth and leadership among participants. These groups often provide training and support to women, equipping them with skills in sustainable practices and entrepreneurship. For instance (Saha, 2011),women involved in crafting and recycling projects learn valuable skills that enhance their employability and income potential. It not only improves their financial stability but also uplift their confidence and social standing within their communities (Sarkar, 2013). The success of these initiatives demonstrates how circular economy practices can serve as a catalyst for women's empowerment and social inclusion.

For example (Saha, 2011),*women's groups involved in waste management and recycling initiatives not only address environmental concerns but also bring community members together. They organize workshops and awareness campaigns to educate people about the importance of waste reduction and recycling. This collective effort enhances community participation and responsibility, leading to improved public health and environmental quality. An illustrative example is shown in Fig. 2.4.*

Fig. 2.4 (Saha, 2011) Remarkable work in waste management and recycling by NGOS for Women in Shillong

2.6.3 Environmental Impact

The environmental impact of women's groups in Shillong, as pioneers of the circular economy, is profound and far-reaching (Saha, 2011). Through their innovative waste reduction, sustainable agriculture, and biodiversity conservation efforts, these groups are driving significant environmental improvements and providing a culture of sustainability (Sarkar, 2013). Their commitment to environmental leadership not only benefits their local communities but also contributes to broader ecological goals (Thapa, 2016). The activities of women's groups in Shillong not only address environmental issues but also enhance community resilience and social cohesion. By engaging in collective environmental actions, these groups build strong social networks and provide a sense of community support. This communal approach is particularly valuable in times of environmental stress or natural disasters, as strong social ties can provide essential support and resources. Moreover, the success of these initiatives often inspires other community members to get involved, further strengthening community bonds and expanding the reach of environmental programs (Sarkar, 2013). The collaborative nature of these efforts demonstrates how environmental sustainability can be integrated with social well-being and community strength (Thapa, 2016). In addition to waste management, women's groups in Shillong are promoting sustainable agriculture and efficient resource use. They advocate for organic farming practices that avoid synthetic chemicals, thus protecting soil health and reducing environmental pollution. By employing natural fertilizers and pest control methods, these women ensure that farming practices are sustainable and environmentally friendly.

For Example, many of these (Saha, 2011) *Women's groups have adopted resource management techniques such as rainwater harvesting and composting. Rainwater*

Fig. 2.5 (Saha, 2011) The women at work in their shed for converting organic waste into nutrient-rich fertilizer, enhancing soil fertility and reducing the need for chemical fertilizers

harvesting helps in conserving water, a precious resource in many parts of India, while composting converts organic waste into nutrient-rich fertilizer, enhancing soil fertility and reducing the need for chemical fertilizers. These practices illustrate the application of circular economy principles in agriculture, promoting environmental sustainability and resilience. An illustrative example is shown in Fig. 2.5.

2.7 Challenges and Solutions

The challenges faced by (Thapa, 2016) women's groups in Shillong in pioneering the circular economy are significant but not unbeatable (Thapa, 2016). Through innovative solutions, strategic partnerships, and unwavering determination, these groups have made remarkable progress in promoting sustainability and driving environmental change. Their success stories serve as powerful examples of how grassroots initiatives can overcome obstacles and create lasting impact (Sarkar, 2013). As they continue to navigate challenges and explore new opportunities, the women of Shillong are paving the way for a more sustainable and equitable future, inspiring communities near and far.

2.7.1 Challenges Faced by Women's Groups

2.7.1.1 Limited Access to Resources and Funding

(Thapa, 2016) Many women's groups in Shillong operate with limited financial resources (Verma, 2021). Access to funding is critical for scaling up projects, purchasing necessary equipment, and covering operational costs. Without adequate funding, these groups struggle to expand their initiatives and impact.

2.7.1.2 Lack of Technical Knowledge and Training

(Verma, 2021) Implementing circular economy practices requires specific technical skills and knowledge (Thapa, 2016). Many women in these (Thapa, 2016) groups lack formal education or training in areas such as waste management, sustainable agriculture, and recycling technologies, which can hinder the effectiveness of their projects.

2.7.1.3 Inadequate Infrastructure

(Verma, 2021) Proper infrastructure is essential for waste collection, recycling, and other circular economy activities (Verma, 2021). In many parts of Shillong, infrastructure is underdeveloped, making it difficult for women's groups to efficiently carry out their work (Verma, 2021). This includes a lack of recycling facilities, transportation for waste materials, and storage spaces.

2.7.1.4 Cultural and Social Barriers

(Wankhar, 2014) Traditional gender roles and societal expectations can pose significant challenges. Women often face resistance when stepping into leadership roles or pursuing entrepreneurial activities. Additionally, balancing household responsibilities with community work can be particularly demanding.

2.7.1.5 Regulatory and Policy Constraints

(Verma, 2021) Navigating the regulatory landscape can be complex. Inadequate or unclear policies related to (Wankhar, 2014) waste management and environmental conservation can create obstacles for women's groups. They may also face bureaucratic hurdles in obtaining permits or accessing government support programs.

2.7.2 Solutions Found by Women's Groups

2.7.2.1 Securing Funding Through Partnerships

To address funding challenges (Women's Environmental Network, 2009), women's groups in Shillong have sought partnerships with NGOs, government agencies, and private sector organizations (Verma, 2021). Grants and microfinance programs have been instrumental in providing the financial support needed to sustain and grow their initiatives. For example (Wankhar, 2014),collaborating with (Women's

Environmental Network, 2009) environmental NGOs has helped secure funding for waste management projects and training programs.

2.7.2.2 Capacity Building and Training Programs

(Women's Environmental Network, 2009) Recognizing the importance of technical skills, many groups have invested in capacity-building initiatives. Workshops and training sessions conducted by experts in waste management (Women's Sustainability Group, 2013), sustainable agriculture, and recycling have equipped women with the knowledge and skills necessary to implement circular economy practices effectively. These programs often focus on practical, hands-on learning, ensuring that participants can immediately apply what they have learned.

2.7.2.3 Innovative Infrastructure Solutions

To overcome infrastructure limitations, women's groups have developed innovative solutions (Women's Waste Management Association, 2015). Mobile recycling units and community-based collection centers have been established to facilitate waste management in areas with inadequate infrastructure (Women's Sustainability Group, 2013). These solutions are often low-cost and designed to be easily replicable, ensuring that they can be implemented in various settings.

2.7.2.4 (Women's Sustainability Group, 2013) Empowerment and Community Engagement

Addressing cultural and social barriers involves empowering women and providing community support (Women's Environmental Network, 2009). Women's groups have launched awareness campaigns to educate communities about the importance of the circular economy and the role of women in driving sustainable change. By showcasing successful projects and highlighting the benefits to the community, they have gained greater acceptance and support for their work.

2.7.2.5 Advocacy and Policy Engagement

(Women's Waste Management Association, 2015) Women's groups have become active advocates for policy changes that support circular economy initiatives. By engaging with local government officials and participating in policy dialogues, they have worked to create a more supportive regulatory environment (Yadav, 2018). Efforts have included advocating for clearer waste management policies, incentives for recycling businesses, and streamlined processes for accessing government support.

Case Study (Mawphor Women's Cooperative, 2020) *One notable example is the Mawphor Women's Cooperative, which has successfully implemented a community-based recycling program. By partnering with local schools and businesses, the cooperative has established a comprehensive waste collection and recycling system. Training programs on waste segregation and recycling techniques have empowered women with the skills needed to manage the program effectively. Additionally, the cooperative has secured funding from environmental NGOs, allowing it to expand its operations and reach more communities.*

2.8 Future Prospects and Sustainability Practices

The concept of the (Yadav, 2018) circular economy (CE) is increasingly recognized as a sustainable alternative to the traditional linear economy (Yadav, 2018). Women's groups in Shillong have embraced the circular economy by implementing innovative strategies to manage waste and promote resource efficiency (Zeliang, 2022). Key practices include waste segregation, recycling, up-cycling, and composting. By transforming waste materials into valuable resources, these groups not only reduce environmental pollution but also create economic opportunities for their communities.

One notable example is the (Mawlynnong Women's Cooperative, 2018) *Mawlynnong Women's Cooperative, which has set a benchmark for waste management in Shillong. The cooperative organizes regular clean-up drives, educates the community on waste segregation, and runs a successful composting program that converts organic waste into nutrient-rich compost for local farmers* (Mawlynnong Women's Cooperative, 2018). *Similarly, the Shillong Handicrafts Collective has excelled in upcycling waste materials into handcrafted products, thereby creating a market for sustainable goods and generating income for women artisans.*

2.8.1 Future Prospects

The future prospects for women's groups in Shillong within the (Zeliang, 2022) circular economy framework is promising. As global awareness of sustainability and environmental conservation grows, there is an increasing demand for (Yadav, 2018) eco-friendly products and services (Zeliang, 2022). Women's groups in Shillong can leverage this trend by scaling up their operations and expanding their market reach. With proper support and investment, these groups can enhance their technical expertise, access better waste management technologies, and improve their product quality (Zeliang, 2022). Collaborations with government agencies, NGOs, and private enterprises can provide the necessary resources and infrastructure to amplify their impact. Moreover, digital platforms and e-commerce can offer new avenues for marketing their products beyond local boundaries, thus ensuring financial sustainability.

(Zoramthanga, 2024) Women's groups in Shillong are at the forefront of the circular economy movement, showcasing how grassroots initiatives can bring significant environmental and socio-economic benefits (Zeliang, 2022). By continuing to innovate and adapt, the groups have the potential to transform Shillong into a model of sustainability. The future prospects are bright, provided they receive adequate support and continue to engage in best sustainability practices (Zoramthanga, 2024). Their journey from waste to wealth not only empowers women but also paves the way for a greener and more resilient future for the entire community.

2.8.2 Sustainability Practices

To ensure long-term (Zoramthanga, 2024) sustainability, women's groups in Shillong adopt comprehensive and adaptive practices. The Key areas of focus include:

(a) *Education and Awareness:* (Zoramthanga, 2024) Continuous education and awareness programs are essential to foster a (Zeliang, 2022) culture of sustainability within the community. By organizing workshops, seminars, and community meetings, women's groups can educate the public on the importance of waste management and sustainable consumption patterns.

(b) *Resource Efficiency:* Implementing resource-efficient practices such as water and energy conservation, as well as the use of (Zeliang, 2022) renewable energy sources, can significantly reduce the environmental footprint of their operations. Adopting a zero-waste approach where possible will further enhance their sustainability credentials.

(c) *Collaboration and Networking*: (Zoramthanga, 2024) Building strong networks with other sustainability-focused organizations can facilitate knowledge sharing and collaborative problem-solving (Zoramthanga, 2024). Participation in regional and national forums can help women's groups stay updated on the latest trends and best practices in circular economy initiatives.

(d) *Monitoring and Evaluation:* Regular monitoring and evaluation of their initiatives can help women's groups assess their impact and identify areas for improvement (Zoramthanga, 2024). Developing metrics to measure environmental, social, and economic outcomes will enable them to demonstrate their contributions to sustainable development convincingly.

2.9 Conclusion

Women's groups in Shillong are leading the charge in pioneering the circular economy, demonstrating that sustainable practices can yield both environmental and economic benefits. Their innovative approaches to (Banerjee, 2010) waste management, resource efficiency, and community education have laid a solid foundation for a more resilient local economy. These initiatives not only address environmental challenges but also create job opportunities, enhance community engagement, and empower women. Looking ahead, the future prospects for these groups are promising. With continued support from government agencies, NGOs, and private sector partners, they can scale up their efforts, adopt advanced technologies, and expand their market reach. Adopting digital platforms and e-commerce can further enhance their financial sustainability and impact. Their dedication and innovative spirit set a powerful example for communities worldwide, proving that grassroots initiatives can bring significant change and pave the way for a sustainable future. For sustained success, the Women's groups must continue to focus on education, resource efficiency, collaboration, and rigorous monitoring of their initiatives. By doing so, they can ensure that their efforts remain effective and impactful in the long term.

References

Banerjee, S. (2010). Role of women in waste management in North-East India. *Journal of Environmental Studies, 23*(2), 45–59.

Baruah, B. (2011). Empowering women through waste management: Case studies from Meghalaya. *International Journal of Social Sciences, 14*(3), 78–92.

Basu, R. (2012). Sustainable development practices in Shillong: Women's contributions. *Eco-Development Journal, 29*(1), 112–125.

Bhowmik, S. (2013). Women and the circular economy: A case study of Shillong. *Journal of Gender and Development, 34*(4), 207–220.

Chhakchhuak, L. (2024, March 2). Garbage to gold: Shillong women's group shows the way. *Citizen Matters*. https://citizenmatters.in/shillong-womens-self-help-group-boosts-waste-recycling/

Choudhury, M. (2014). Waste management initiatives by women in Meghalaya. *Journal of Urban Studies, 45*(3), 145–160.

Das, P. (2015). The impact of women's groups on recycling practices in Shillong. *Journal of Environmental Economics, 39*(2), 85–99.

Deka, A. (2016). Grassroots innovations in waste management: The Shillong experience. *International Journal of Environmental Studies, 21*(1), 95–110.

Ghosh, S. (2018). The role of women's cooperatives in promoting circular economy in Shillong. *Journal of Community Development, 37*(3), 120–135.

Gogoi, R. (2019). Waste to wealth: Case studies from women entrepreneurs in Shillong. *Journal of Environmental Innovation, 42*(4), 150–165.

Hazarika, N. (2020). Women and sustainability: The Shillong model. *Journal of Rural Development, 45*(2), 90–105.

Kharmawphlang, P. (2021). Circular economy practices by women's groups in Meghalaya. *Journal of Environmental Management, 50*(1), 60–75.

Mawlynnong Women's Cooperative. (2018). Sustainability initiatives in Mawlynnong village. *Eco-Friendly Practices, 30*(3), 210–225.

Mawphor Women's Cooperative. (2020). Community-based recycling programs in Meghalaya. *Journal of Community Recycling, 34*(2), 95–110.

Mishra, A. (2011). Women's role in waste management: A study in Shillong. *Journal of Environmental Research, 26*(1), 65–80.

Nongkynrih, W. (2012). The impact of women's initiatives on waste reduction in Shillong. *Journal of Social Impact, 29*(2), 85–100.

Qabeer, V. a. P. B. (2018, February 28). The journey of Shillong's waste-pickers to become compost suppliers to Meghalaya Raj Bhawan. Stories of waste and waste workers – live blog of Hasiru Dala. https://wastenarratives.com/2018/02/28/the-journey-of-shillongs-waste-pickers-to-become-compost-suppliers-to-meghalaya-raj-bhawan/

Saha, D. (2011). The economic impact of waste management by women in Shillong. *Journal of Economic Studies, 32*(2), 115–130.

Sarkar, S. (2013). Women and the circular economy: Lessons from Shillong. *International Journal of Sustainability, 25*(3), 160–175.

Thapa, P. (2016). Empowering women through sustainable waste management in Shillong. *Journal of Community Empowerment, 31*(1), 95–110.

Verma, S. (2021). Women in the circular economy: The case of Shillong. *Journal of Gender and Environmental Studies, 48*(1), 115–130.

Wankhar, E. (2014). The role of women in promoting sustainability in Shillong. *Journal of Green Practices, 35*(3), 85–100.

Women's Environmental Network. (2009). Women's role in environmental conservation in North-East India. *Journal of Eco-Conservation, 17*(2), 90–105.

Women's Sustainability Group. (2013). Best practices in waste management by women's groups in Shillong. *Journal of Waste Management, 27*(3), 115–130.

Women's Waste Management Association. (2015). Innovative waste management by women in Meghalaya. *Journal of Environmental Innovation, 39*(2), 125–140.

Yadav, K. (2018). Circular economy initiatives by women's groups in Shillong. *Journal of Eco-Innovation, 42*(3), 105–120.

Zeliang, P. (2022). Women's contributions to the circular economy in North-East India. *Journal of Environmental Economics, 47*(2), 140–155.

Zoramthanga, M. (2024). Women empowerment through waste management in Shillong. *Journal of Sustainable Development, 53*(1), 130–145.

Chapter 3
Transforming Waste to Wealth Leading to SDG of United Nations: A case Study from ITI Berhampur, India

Rajat Kumar Panigrahi

Abstract Traditionally, TVET curriculums have focused on a combination of knowledge, technical skills, and generic skills. However, the industrial landscape is evolving. Today, there's a growing demand for a new kind of workforce: one competent in green skills. In a unique effort promoting sustainable development through education and green skill, Government ITI Berhampur has empowered its trainees to design and fabricate beautiful sculptures from waste materials generated in the institute's workshops during training practice. Waste materials are being given a new life by transforming it into aesthetically pleasing sculptures. This initiative not only enhances hands-on skills like fitting, welding, painting, but the finished products are also in high demand by interior decorators, creating a potential income stream for the trainees. This program effectively combines environmental consciousness aligning with sustainable development goals of United Nations turning the trainees into sustainability warriors while transforming the campus into an Educational Wonderland which has etched its name in the illustrious Asia Book of Records. This innovative approach is fostering a keen interest among students in careers with the Technical and Vocational Education and Training (TVET) sector. This is leading to a rise in female participation, increased employability with better salaries, a minimized skill gap and attracts more students for admissions, ensuring a future generation equipped for the green economy.

Keywords Green skill · Waste to wealth · Up skilling · Waste management · Educational wonderland

R. K. Panigrahi (✉)
Skill Development and Technical Education and Training, Govt Industrial Training Institute, Berhampur, India

S. Patnaik et al. (eds.), *Circular Economy and Sustainable Value Creation through Eco-Innovation*, Circular Economy and Sustainability,
https://doi.org/10.1007/978-3-031-88516-7_3

3.1 Introduction

Technical and Vocational Education and Training (TVET) institutions often face an uphill battle. Despite equipping students with valuable skills, they're frequently seen as "second-tier" or "second-chance" schools. This perception leads to the belief that TVET graduates are less competitive than university graduates, destined for lower-paying, less prestigious jobs. This creates a self-fulfilling cycle: stronger students, often from higher socioeconomic backgrounds, are encouraged towards universities, while those perceived as "less successful"—frequently from lower socioeconomic classes—are directed into TVETs. This reinforces the negative image of TVETs, hindering their ability to attract a broader range of students and fully contribute to a skilled workforce (Legusov et al., 2021).

Additionally TVET face a number of hurdles. One key challenge is attracting female participation. Societal norms and a lack of targeted programs can discourage women from pursuing vocational careers. Graduates may face difficulties securing employment due to a skills gap or a perception that TVET qualifications hold less weight than university degrees (Raby et al., 2022). This, in turn, can limit salary growth potential. Lagging in hands-on skill can also be an issue, with curriculums struggling to keep pace with rapidly evolving industry demands. Finally, inadequate learning facilities, such as outdated equipment or insufficient resources, can hinder the quality of training provided. These combined challenges make it difficult for TVET programs to deliver on their full potential of creating a skilled and competitive workforce (Fig. 3.1).

India's ITI network, the backbone of vocational training, faces a complex challenge. Despite numerous studies and recommendations by government bodies, issues persist. NITI Aayog reports highlight low-capacity utilization, with only 10.5 lakh trainees filling the 25 lakh available seats. Even more concerning are the dismal placement rates, reflecting a wider problem with the ITI ecosystem. The

Fig. 3.1 Challenges in TVET

quality of training and societal perception of ITIs remains low, creating a vicious cycle that discourages enrolment and investment. This situation demands a renewed focus on improving ITI programs to meet the skill gap of India's workforce.

Moreover Technical and vocational education (TVET) plays a vital role in bridging the knowledge and skills gap critical to achieving the Sustainable Development Goals framed by United Nations. Beyond just technical skills, well-designed TVET programs can also instil the right mindset and attitudes within trainees, shaping a future workforce that embraces sustainability. To maximize this impact, however, strengthening relevant policies and skills development strategies is essential (Mohamed & Wahba, 2011).

The "Green Skills for Waste to Wealth" initiative at ITI Berhampur in Odisha, India, is a paradigm shift in technical education (Kalkanis et al., 2022). It seamlessly integrates skill development with responsible waste management practices, tackling two critical issues simultaneously: the burgeoning skill gap in the green sector and the rising tide of waste generation within the institute itself. This program empowers the students with valuable hands-on skills in the circular economy's "waste to wealth" approach. Unlike traditional classroom learning, hands-on skills are honed through active engagement and practical application. Students can immediately practice what they learn, progressively enhancing their abilities and minimizing the industry-readiness gap. This initiative not only prepares them for potential careers but also opens doors for entrepreneurial ventures. Figure 3.2 by focusing on innovative waste-to-wealth methods, the program fosters responsible waste management and a positive carbon footprint, promoting a more sustainable future (Panigrahi & Patnaik, 2023).

Govt. ITI Berhampur, established in 1957, is a premier institution for vocational training in India (Fig. 3.3). It is the largest ITI in the state and sixth largest ITI in the country offering a diverse range of long-term and short-term training programs for skill development with industry- relevant skills and knowledge to over 4000 aspiring individuals in 27 sectors (Fig. 3.3).

Our students are leading the charge in sustainable practices by adopting innovative up cycling techniques. Their approach transforms discarded materials like scrap metal, e-waste, electrical components, PVC wires, automobile scraps, iron waste from demolished buildings (Bsisu & Salem, 2020) and even used tyre from vehicles

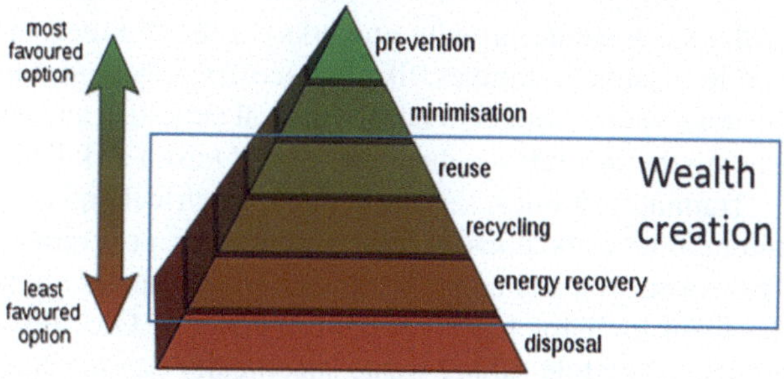

Fig. 3.2 Waste Hierarchy Pyramid

Fig. 3.3 Govt ITI Berhampur, Odisha

into valuable and beautiful sculptures. This initiative achieves a zero environmental impact, perfectly aligning with the growing emphasis on Green TVET.

Green TVET (Xaime et al., 2020) prioritizes creating a clean and sustainable future. ITI Berhampur's creative approach exemplifies how TVET institutions can address this crucial challenge. By ingeniously reusing waste materials and transforming them through their existing skills, the students are not only developing their technical abilities but also fostering a more responsible and environmentally conscious mindset. This innovative program paves the way for a greener future in technical education.

Developing the right skill set is crucial for TVET graduates to find success in today's job market. To bridge the gap between industry needs and ITI training, a focus on hands-on experience is essential. Here, the innovative design thinking lab at ITI Berhampur plays a vital role. The lab equips students with practical skills by guiding them in creating contemporary sculptures suited for the booming interior design industry (Fig. 3.4).

By equipping students with industry-relevant technical knowledge and practical skills, the institute proudly embarks on the "Skilled in Odisha" vision (Fig. 3.5). The innovative initiatives, like the eco-friendly sculptures made from waste, demonstrate a commitment to both technical expertise and environmental responsibility, making the institute a valuable asset to Odisha's skilled workforce development.

Training institutes, service centres, electrical and electronic equipment labs, and automobile workshops grapple with waste generation during training exercises. This waste typically consists of materials discarded after students' initial use, deeming them unfit for further training purposes. Globally, material waste constitutes around 4%, while rubber waste specifically can reach up to 17% of the total waste composition (Fig. 3.6).

Fig. 3.4 Hands-on training

Fig. 3.5 Skilled in Odisha

While recycling metal scraps like TMT bars can contribute to global warming (2.1 kg CO2eq/kg HRC), up cycling them offers a significant environmental benefit. Up cycling these bars, often collected from demolished buildings, prevents the CO2 emissions associated with re-melting them (Suer et al., 2022). Additionally, their inherent strength makes them ideal for reuse in construction projects, as long as their bond and tensile strength haven't degraded.

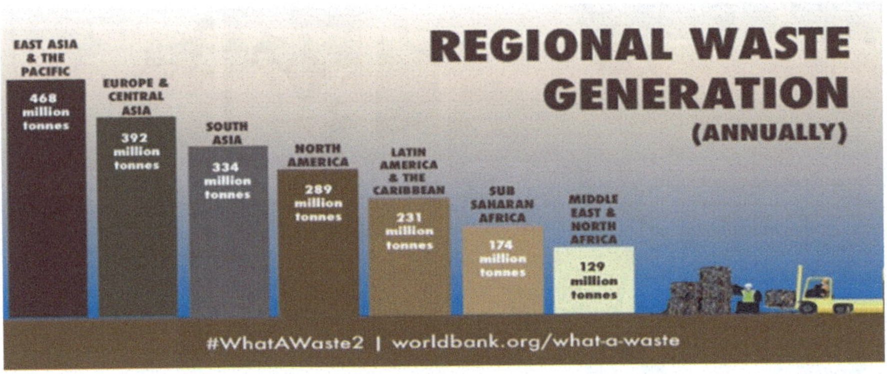

Fig. 3.6 Global waste generation

On the other hand, scrap tires in city yards pose a serious environmental threat. Exposed sunlight degrades them, contaminating soil and potentially groundwater. Additionally, they can catch fire, releasing harmful pollutants into the air. These discarded tires also create breeding grounds for mosquitoes, spreading diseases like dengue.

Thankfully, up cycling offers a solution. By transforming these scraps into useful objects, ITI Berhampur students are tackling this issue head-on. This initiative showcases the power of up cycling in reducing waste and promoting a cleaner environment leading to circular economy (Wali et al., 2019) and SDGs of United Nations garnering national and international recognition.

The 5 feet Mahatma Gandhi and 8 feet Fish sculpture designed and fabricated of wasted fan bearings using SMAW welding aligns with United Nations SDG 16: Peace, Justice And Strong Institutions and SDG 14: Life Below Water awarded as winner and Runners Up in Welded Marvels “Project Trash to Treasure” organized by Indian Institute of Welding, India for the year 2022 respectively (Fig. 3.7). The 6 feet Konark Chakra sculpture designed and fabricated of wasted fan bearings and bike cranks aligns with United Nations SDG 8: Decent Work and Economic Growth awarded as Runners Up in Welded Marvels “Project Trash to Treasure” organized by Indian Institute of Welding, India for the year 2023 respectively (Fig. 3.8).

The 23 feet tall elephant made from 30,000 scrap plastic bottles on the occasion of cleanliness day. The project aligns with United Nation’s SDG 12: Responsible Consumption and Production. By reusing plastic bottles instead of discarding them, ITI Berhampur promotes responsible waste management and reduces the environmental impact of plastic pollution (Fig. 3.9).

These beautiful, up cycled art pieces not only enhance the students’ hands-on skills but also foster self-sustainability. By utilizing readily available waste materials, students generate revenue during their training, preparing them for a future where “waste to wealth” products are increasingly sought after. The interior decoration market itself is projected to reach USD 58.24 billion by 2029 (Fig. 3.10). This trend aligns perfectly with the institute’s focus on up cycled sculptures, making the approach not just innovative but highly relevant to the evolving industry landscape.

Fig. 3.7 Achievements in Welded Marvels competition

Fig. 3.8 Konark Chakra sculpture

Fig. 3.9 Elephant sculpture from scrap plastic bottles

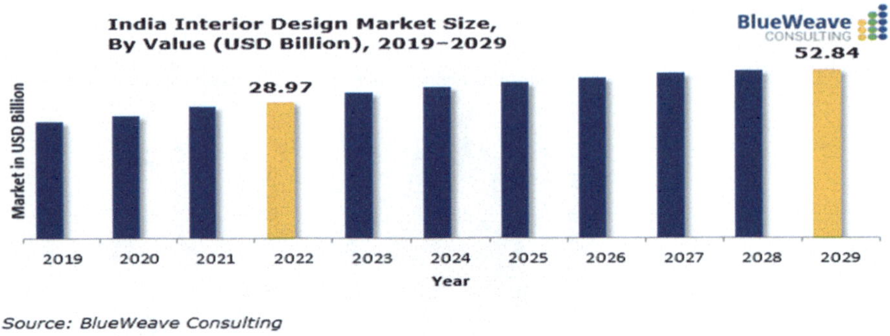

Fig. 3.10 Interior decoration industry evolution

The G20 recognizes the importance of a skilled workforce, adaptable to future needs, for national and global growth. They emphasize collaboration to address this challenge. A key concern is the need for continuous skill development and reskilling to keep pace with technological advancements. India, with its young population, is well-positioned to capitalize on this demand by providing a skilled workforce to the world. However, ensuring these young people have the necessary industry-specific skills remains a challenge. This is where initiatives like ITI Berhampur's waste-to-wealth program come in. By equipping trainees with practical skills in waste management and up cycling, they bridge the skill gap and prepare graduates for the demands of the job market.

3.2 Methodology

Technical institutes like ITI Berhampur in India grapple with a significant amount of waste generation. With over 4000 students and multiple trades, the institute produces roughly 1000 kg of scrap every month (Fig. 3.11). This include metal scraps, electronic components like PCBs, discarded spare parts from vehicles, and even iron dust that pollutes the soil and water. Traditional disposal methods like landfills are unsustainable, prompting ITI Berhampur to adopt a groundbreaking approach—integrating skill development with waste management.

This initiative tackles multiple challenges simultaneously. It addresses the skill gap by providing students with practical experience in waste management and up cycling techniques. This prepares them for industry demands and potentially entrepreneurial ventures. Additionally, by reusing scrap materials, the program minimizes the institute's carbon footprint and promotes responsible waste management.

The resourceful program collects various materials for up cycling. Scrap iron bars are straightened and repurposed. Discarded metal from window doors and water tanks from government departments provides additional resources. Automobile workshops within the city and the institute's own scrap yard contribute

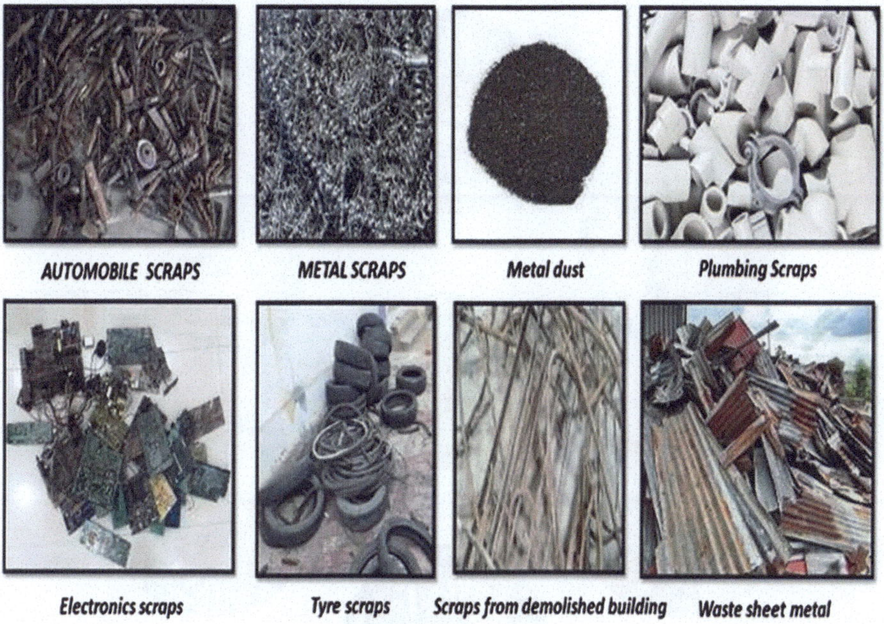

Fig. 3.11 Type of scrap materials generated

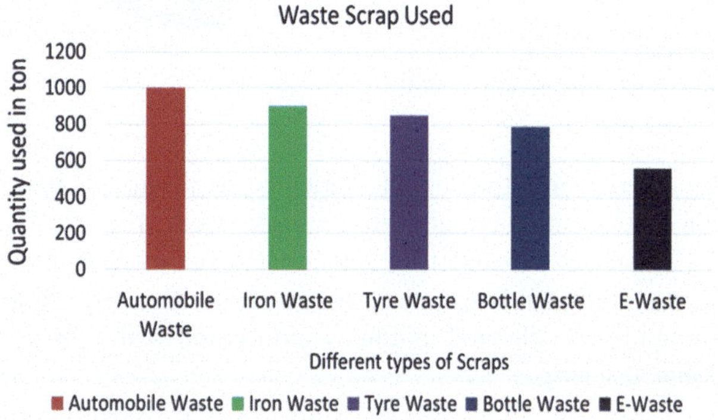

Fig. 3.12 Types of scrap materials used

gears, wheels, and other vehicle parts. E-waste, a major environmental concern, is collected from both the institute and local collection centers. Even plastic bottle caps from the campus canteen are not spared, finding new life in innovative sculptures. Finally, the ever-growing pile of discarded (Turer, 2012) from cars, scooters, and other vehicles find a creative outlet in the program (Fig. 3.12).

The design thinking lab orchestrates the transformation. Here, 3D models of sculptures are meticulously planned using computer software. These plans are then translated into large-scale drawings on the workshop floor. A team of trainees from various trades—welding, painting, fitting, and CNC turning—comes together to

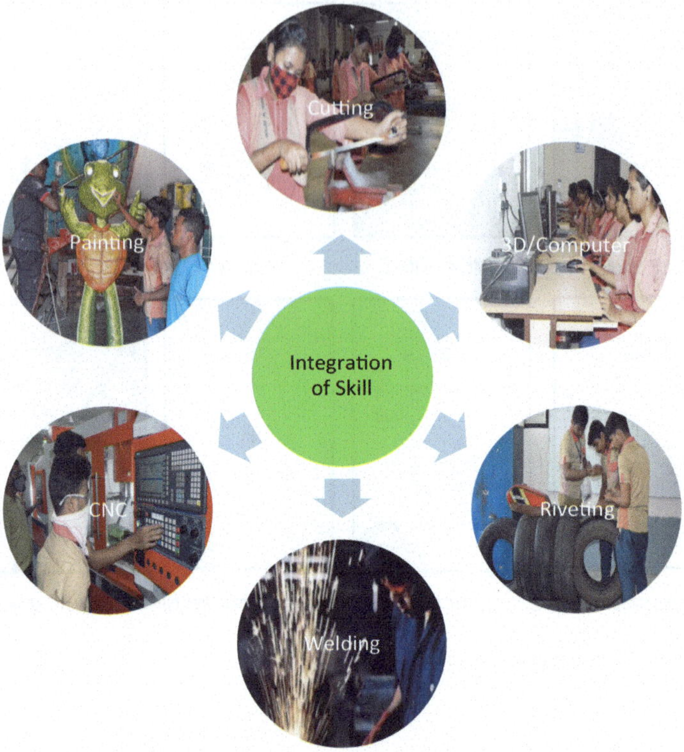

Fig. 3.13 Integration of skill

breathe life into the designs. Industry professionals provide valuable technical guidance, fostering a collaborative environment (Fig. 3.13).

The final product is a testament to teamwork, as students utilize their hands, hearts, and minds to create aesthetically pleasing sculptures. These sculptures are meticulously cleaned, painted, and treated with a protective coating to ensure their longevity (Fig. 3.14).

These sculptures not only showcase the students' newfound skills but also serve as a powerful symbol of ITI Berhampur's commitment to a sustainable future. This initiative demonstrates how waste management and skill development can be effectively integrated to create a win-win situation for both the environment and the workforce. This not only provides students with valuable experience but also fulfills the growing demand for sustainable art in sectors like hotel and corporate interior design, park beautification, fashion shows as thematic elements, and even residential decor. Theme parks established by governments and NGOs are also showing significant interest in these unique sculptures.

This collaborative effort, involving various trades within the institute, has yielded impressive results. The initial success culminated in the creation of a dedicated

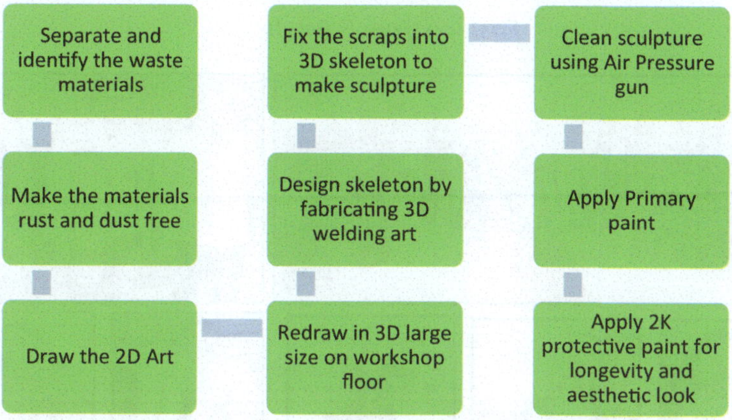

Fig. 3.14 Methodology

Fig. 3.15 Waste to Wealth Park at ITI Berhampur

"Waste to Wealth Park," showcasing the unique sculptures from discarded materials like e-waste, tires, and automobile parts as a testament to the students' artistry and the program's effectiveness (Fig. 3.15). This park has become a magnet, attracting visitors, interior designers from across regions, and demonstrating the immense potential of up cycled art in the flourishing interior decoration market.

Government ITI Berhampur's "Waste to Wealth" program isn't just about creating sculptures from discarded materials; it's also a platform for sustainable fashion innovation. The fashion show features one-of-a-kind outfits crafted entirely from e-waste! Students fabricate old keyboards, capacitors, motherboards, and PCBs into dresses, jackets, shoes, and purses (Fig. 3.16).

To cater to a wider audience, ITI Berhampur's waste-to-wealth initiative offers sculptures in three distinct size categories: Small, intricate pieces (under 5 kg) are perfect for homes and can be customized to your taste. Priced up to ₹30,000

Fig. 3.16 Fashion show from E-waste

Fig. 3.17 Artifacts using Bottle caps and E-waste

internationally, they're a unique and affordable option. Medium sculptures (40-50 Kg) make a statement in offices, reaching up to ₹50,000 in value. Finally, large-scale pieces (up to 5 tons) become city landmarks, potentially generating revenue for the program and fetching up to USD 20,000 in the international market. No matter the size, these sustainable and artistic creations are a great investment (Fig. 3.17).

Waste management is a growing concern globally, and the G20 has recognized the importance of promoting sustainable waste management practices. By participating in G20 and displaying these sculptures, ITI Berhampur is likely raising awareness about the importance of waste reduction and promoting creative solutions to environmental problems (Fig. 3.18).

Fig. 3.18 Participation in G20

Fig. 3.19 Selfies with Mascot in General Election 2024

ITI Berhampur created the eco-friendly sculptures from waste materials to use at a polling booth in the 2024 general election. The Berhampur Municipal Corporation (BeMC) commissioner visited the polling station and was very impressed. Voters were very excited to take selfies with the sculptures (Fig. 3.19.

The sculptures included a map of India made from e-waste, a turtle made from bottle caps, and chairs and EVM stands made from tires (Fig. 3.20).

Symposiums have been organized to promote these green skills around the community (Fig. 3.21).

To make these products available to the public, 24/7 waste to wealth outlet has been opened in the ITI campus. The outlet accepts cashless transactions for a smooth buying experience (Fig. 3.22).

Fig. 3.20 Eco Friendly Polling Booth from waste materials for General Election-2024

Fig. 3.21 Symposium on Waste to Wealth

Fig. 3.22 Waste to wealth outlet

3.3 Results and Discussion

This innovative integration of skill development and waste management offers a double benefit. Firstly, trainees gain essential hands-on experience in areas like waste management and up cycling techniques. This practical learning bridges the skill gap between what they learn and what industries demand, making them much more prepared for the workforce. Secondly, the design thinking lab serves as a valuable tool for honing vital soft skills. By collaborating on real-world projects like these sculptures, trainees develop their teamwork abilities, communication skills, and creative problem-solving approaches. This fosters adaptability and entrepreneurial thinking, equipping them for success not just in traditional jobs but also in potential future ventures.

The initiative has yielded impressive results beyond just creating art.

The institute's internal revenue has grown steadily, likely due to sales of the sculptures (Table 3.1).

More importantly, the program has significantly reduced the skill gap, leading to a placement increase from 602 to 1425 students in 5 years (Table 3.2).

This translates to a rise in average student salaries, jumping from ₹11,500 to ₹22,800 during the same period (Table 3.3).

The program's success has also attracted more applicants, with admissions tripling from 3000 to 11,000 in just 3 years (Table 3.4).

Notably, there's been a surge in female enrollment, rising from 427 to 734 in the last 3 years (Table 3.5).

Table 3.1 Bar chart showing internal revenue generation in lakh over three years. In 2021, revenue is 18.78 lakh; in 2022, it increases to 19.12 lakh; and in 2023, it further rises to 20.87 lakh

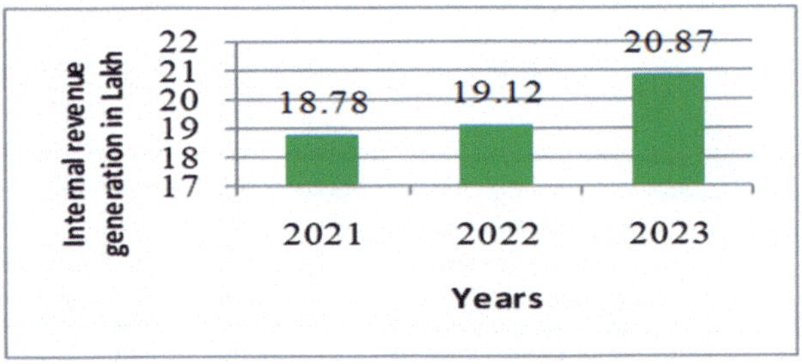

Table 3.2 Bar chart showing the number of students placed from 2019 to 2023. The placements increase each year: 602 in 2019, 1008 in 2020, 1039 in 2021, 1131 in 2022, and 1425 in 2023. The chart highlights a consistent upward trend in student placements over the five-year period

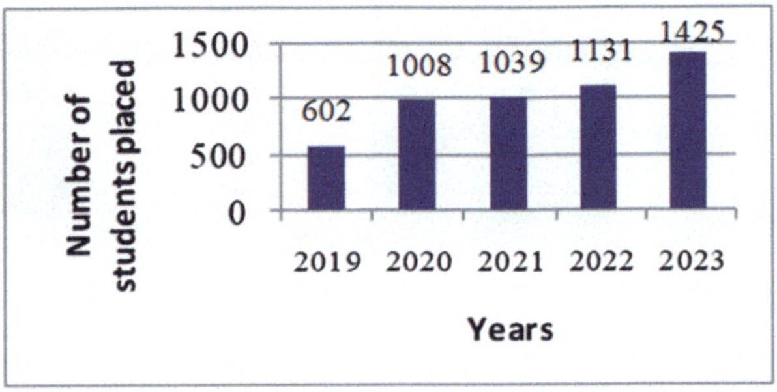

Table 3.3 Bar chart showing the average salary in rupees from 2019 to 2023. The salary increases each year: 11,500 in 2019, 13,350 in 2020, 20,600 in 2021, 21,400 in 2022, and 22,800 in 2023. The bars are green, indicating a positive trend over the years

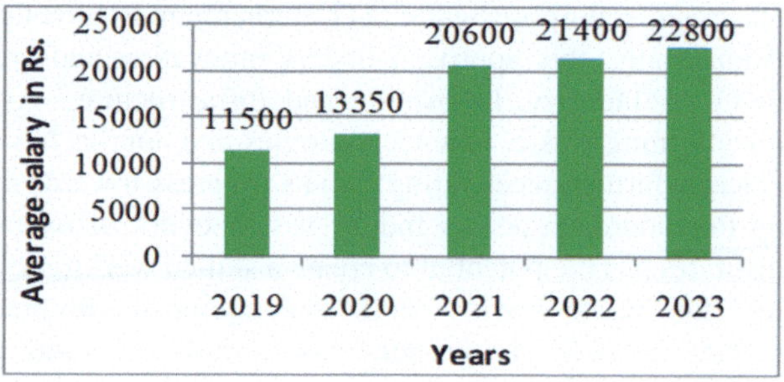

Table 3.4 Bar chart showing the number of applications received over three academic years. In 2021-22, 3,000 applications were received. This increased to 5,000 in 2022-23, and further rose to 11,000 in 2023-24. The chart indicates a significant upward trend in application numbers over the period

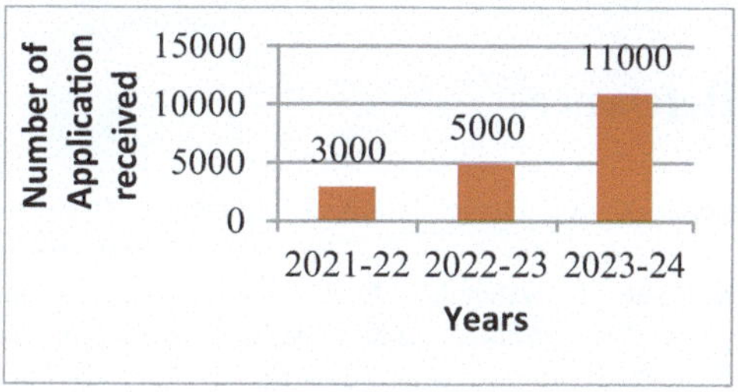

Table 3.5 Bar chart showing the number of girls admitted over three academic years. In 2021-22, 427 girls were admitted; in 2022- 23, 430; and in 2023-24, the number increased to 734. The vertical axis represents the number of girls admitted, and the horizontal axis shows the academic years

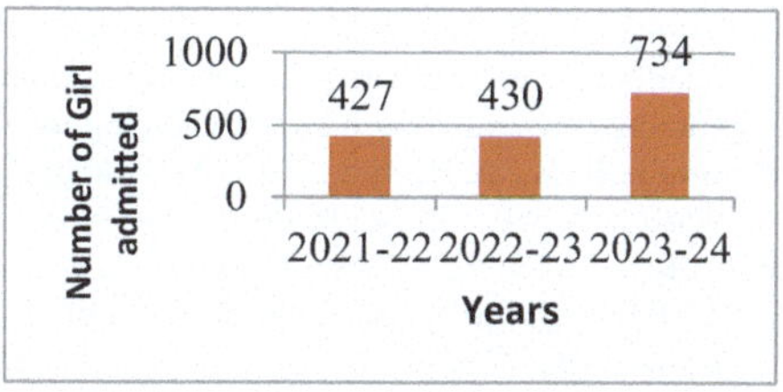

This innovative project not only promotes sustainability but also empowers students with valuable skills, leading to a brighter future for both the institute and its graduates.

3.4 Conclusion

Empowering trainees with the "waste to wealth" mindset is a strategic investment in India's future. By integrating this concept throughout their vocational training at institutes like ITI Berhampur, a generation of skilled professionals can emerge, adept at transforming waste into valuable resources. Integrating the "waste to wealth" concept aligns perfectly with the United Nation's SDGs. SDG 12, "Responsible Consumption and Production," emphasizes the need to manage waste streams sustainably and minimize environmental impact. By training individuals to reuse and repurpose waste, ITI Berhampur is directly contributing to this goal. Additionally, this approach fosters innovation and creativity, which aligns with SDG 9, "Industry, Innovation and Infrastructure." As trainees develop skills in transforming waste, new technologies and approaches to waste management may emerge, further accelerating India's progress towards a sustainable future. Scaling up this approach across India's vast network of over 15,000 vocational training institutes has the potential to create a skilled workforce that champions sustainability. This "waste warrior" workforce, equipped with practical skills and a problem-solving mindset, can propel India towards becoming a global leader in the skill development arena.

References

Bsisu, K. Al.-D., & Salem, Z. A. (2020). Recycling of steel bars from demolished structures. *International Journal of Engineering Research and Technology, 13*(1), 94–99.

Kalkanis, K., Alexakis, D. E., Kyriakis, E., Kiskira, K., Lorenzo-Llanes, J., Themelis, N. J., & Psomopoulos, C. S. (2022). Transforming waste to wealth, achieving circular economy. *Circular Economy and Sustainability, 2*, 1541–1559.

Legusov, O., Raby, R. L., MoU, L., & Gomez-Gajardo, F. (2021). How community colleges and other TVET institutions contribute to the united nations sustainable development goals. *Journal of Further and Higher Education, 46*(1), 1–18.

Mohamed M. & Wahba M. (2011). Technical and Vocational Education and Training (TVET) challenges and priorities in developing countries. https://unevoc.unesco.org/e-forum/TVET_Challenges_and_Priorities_in_Developing_Countries.pdf.

Panigrahi, R. K., & Patnaik, P. (2023). Integration of skill development & waste management for waste to wealth. In *International Research Symposium*.

Raby, R. L., Mou, L., Gómez-Gajardo, F., & Zhou, Y. (2022). How community colleges and other TVET institutions contribute to the united nations sustainable development goals. *Journal of Further and Higher Education, 46*(1), 89–106.

Suer, J., Ahrenhold, F., & Traverso, M. (2022). Carbon footprint and energy transformation analysis of steel produced via a direct reduction plant with an integrated electric melting unit. *Journal of Sustainable Metallurgy, 8*, 1532–1545.

Turer, A. (2012). Recycling of scrap tires. In *Material recycling—Trends and perspectives* (pp. 195–214).

Wali, E., Wizor Collins, H., & Nwankwoala, H. O. (2019). Waste-to-wealth, towards a sustainable zero-waste in a circular economy: An overview. *International Journal of Emerging Engineering Research and Technology, 7*(11), 1–11.

Xaime, A. F., Othman, N. M., Sern, L. C., Hasan, A., & Said, S. M. (2020). A green skills framework for TVET curricula. *International Journal of Innovative Technology and Exploring Engineering (IJITEE), 9*(3), 2532–2534.

Webpage

https://www.business-standard.com/article/economy-policy/india-launches-mission-to-skill-400-million-by-2022-115071600035_1.html
https://www.businesstoday.in/current/corporate/urbanclap-joins-modi-govt-skill-india-mission-train-30000-professionals/story/364467.html
https://www.entrepreneur.com/article/327714
https://www.blueweaveconsulting.com/report/india-interior-design-market
https://ncvtmis.gov.in/Pages/ITI/Search.aspx

Chapter 4
The Contribution of the Circular Economy to the Demand for Socio-Environmental Rights of a Territory: A Reflection Between Utilitarian and Ontological Reason

Francesco Vigliarolo

Abstract The chapter aims to state that the circular economy contributes to the demand for socio-environmental rights of people, without which the human and natural life of a society would be in danger. In order to address this statement, a historical reconstruction is proposed with respect to the importance of the environmental issue that arose in the last century, highlighting the problems that the dominant economic models are creating (social and environmental); subsequently, some concepts are introduced to address the arguments, such as a phenomenological approach in the light of which the demand for rights and the ontological reason of territories are defined, in opposition to the demand for consumption and individual interests. After defining some characteristics of the circular economy, some passages of economic thought are exposed to outline the tension between circular and linear economy, between utilitarian reason and ontological reason. The physiocratic school and some limits of economic thought are drawn, taking up Smith and Keynes.

Finally, concepts such as meso-economy are proposed, as a space where the circular economy could position itself to permanently build with more coherence an ontological reason linked to local identities, and to carry forward the demand for socio-environmental rights of a people. For this reason, it is stated that a circular economy is stronger when there is a change of consciousness, human and social, that considers such practices important for their full individuality and happiness (referring to the circular economy). For these reasons it is argued that in economic thought, the circular economy is positioned as antithetical to the classical economy, based on individual interests and the concept of infinite growth. It reintroduces

F. Vigliarolo (✉)
National University of La Plata, La Plata, Argentina

Catholic University of La Plata, La Plata, Argentina

University of Buenos Aires, Buenos Aires, Argentina
e-mail: francesco.vigliarolo@presi.unlp.edu.ar

S. Patnaik et al. (eds.), *Circular Economy and Sustainable Value Creation through Eco-Innovation*, Circular Economy and Sustainability,
https://doi.org/10.1007/978-3-031-88516-7_4

principles such as circularity and reciprocity that were also proposed several centuries ago.

Keywords Circular economy · Social crisis · Environmental crisis · Sustainable development · Self-sustainable development · Reciprocity · Civil economy · Demand for right · Ontological reason · Utilitarian reason · Individual interest · Financialization · Phenomenology economy · Generative economy · Meso-economy · Physiocrats · Smith · Keynes · Genovesi · Rio de Janeiro 1992

4.1 Introduction

The concept of circular economy that we know nowadays and is establishing in economic thought around the world, in our opinion, is undoubtedly due to various causes, but in this text we would like to highlight at least two:

– the first, on a practical level, refers to the serious environmental problems that have brought to the world's attention the need to find ways of recycling materials to mitigate the negative impact of the market economy (externalities). Over the years, the classical economy, also based on the unlimited hoarding of resources, has transformed societies and the world into a battlefield where the same resources are contested as power games, putting at risk the possibility of exercising human rights for all the inhabitants of the planet, even within the rich countries themselves.
– the second, on the theoretical level, concerns the fact that since Smith, the infinite and linear growth models have progressively led to the loss of the territory-society-economy relationship, as a relationship of reciprocity. Today this break is witnessed by the process of economic financialization that has pushed to exacerbate the creation of wealth without relation to the socio-environmental identity of the populations, what we will call in these pages ontological reason. Consider that only 95% of financial values have a relation to the goods and services exchanged.

The first category of causes has led us to what is known as the environmental crisis but also the social crisis. From the 80s onwards, in fact, new development models have been born that try to mend the relationship with the environment; we first talk about compatible development, then sustainable and from the 90s onwards self-sustainable, where social participation becomes essential. This category undoubtedly includes the crisis of the concept of post-war economic development, and the awakening of a new awareness that has led to the birth of new responses at a conceptual and practical level. The second refers to the limits of the dominant economic theory, based on individual freedoms and the existence of a linear economic model that presupposes infinite growth in a finite world. It is in this context, in fact, that new economic theories such as social, civil, solidarity economy, etc. are born or are taken up again, which propose themselves as a third sector, neither State nor market, in the economic system. These are socially oriented economies that have had exponential growth worldwide since the 1990s compared to the traditional economy.

Following, we will try to analyze the circular economy starting from these categories of causes that are obviously connected to each other. We will retrace some historical stages of the concept of sustainable development, introducing the concept of reciprocity, from Genovesi to the present day, dialoguing with the main schools of economic thought, Smith and Keynes, to introduce a vision of the circular economy linked to the ontological reason of society. To this end, some concepts are proposed such as a circularity not only due to the use of materials, but also of ideas that the economy itself is called to permanently generate by dialoguing with the territory. To this end, we propose a defined phenomenological approach with which we define a generative circular economy, defined as such when it is capable of permanently generating the social and environmental fabric in terms of rights, or demand for socio-environmental rights.

4.2 Some Historical Stages of a New Concept of Development

4.2.1 Environmental Crisis and the Affirmation of the Concept of Sustainable Development

The environmental crisis is generally traced back to the late seventies of the last century, when the first movements and World Conferences on the Environment were born, for example Stockholm 1972. However, long before these dates, a physicist, Svante Arrhenius, Nobel Prize for Physics in 1903, was the first scientist to mention global warming in 1896, proposing a relationship between the concentration of carbon dioxide and atmospheric temperature. But it took almost a century to be heard.

A watershed date is certainly 1992 when the Conference on Environment and Development was held in Rio de Janeiro, that is, for the first time at a global level, an attempt was made to link development starting from the environment. In these years, the concept of sustainable development was born and developed, to be precise in 1987, with the Brundtland Report of the United Nations, which defined it as "the development that allows us to satisfy the needs of the present world without compromising the needs of future generations".

In Rio de Janeiro, important concepts such as "participation", "quality of interventions", "integrated systems", "social responsibility" and other related concepts are proposed based on the involvement and active contribution of civil society. Attention is focused on the social process at the base of development. In this direction, in 1994, in Alborg, the homonymous charter was signed, where 80 European cities and 253 representatives of international organizations adhered to the "I

European Conference of Sustainable Cities[1]". The charter commits the city administration to implement long-term local action plans aimed at improving the local community, and to promote the identification of problems and their causes through public consultation. In August 1994, the International Conference on Population and Development (ICPD) was held in Cairo, where most developing countries committed themselves to implementing active family policies, highlighting the need to review the position of women, especially in terms of education and economic independence. In 1996, the "Istanbul Declaration" and the "Habitat II Agenda" were announced, announcing the birth of the so-called good practices, which consist of the adoption of projects aimed at sustainable development such as: energy savings in homes, integrated management of cycles which encourages the reduction of the use of raw materials, recycling of materials and reuse of land resources and reduction of environmental impact, among others. In December 1997, the "Third Conference of Contracting States" was held in Kyoto, which gave rise to the eponymous protocol. There, the proposal to reduce the use of gas, which causes the greenhouse effect, was approved, where it was proposed to reduce the global average of 1990 of 5.2% between 2008 and 2012. Japan, the USA, and the European Union were to progressively reduce their emissions by 6%, 7%, and 8% respectively.

Since the Bruntland Report, there have been different approaches to sustainable development. Those that claim to have a scientific basis are those that define themselves as economic and territorialist approaches. The former includes conceptualizations that address sustainability as a solution to the problems induced by the economy within the economy and are distinguished as: functional, environmental and ecological. The territorialist approach speaks of self-sustainability and puts the action of organized groups at a local level at the center.

The functionalist approach envisages environmental policies as a specific branch of economics, anchored in the traditional laws of production and consumption. Development must be "compatible" with the requirements of survival and the reproduction of the "physical base" (natural and human). The solution to environmental problems is provided by the self-regulating market, with a series of corrections and procedures that put a "limit" on the causes of pollution and the exploitation of resources beyond the necessary reproduction of the same. The essence of this approach lies in the necessary corrections (limits) and in the concept of "carrying capacity of ecosystems" which are the framework within which development must be contained in order to be "compatible". The main criticisms are related to the fact that:

- interventions are aimed at reducing negative impacts on the environment and not at definitively eliminating the causes that provoke such a negative result;
- the fundamental role is played by the market, which intervenes only in monetary goods, excluding any other socio-environmental value in economic activities;

[1]The Alborg conference was organized in collaboration with ICLEI (International Council for Local Environmental Initiatives), an international organization that began its activity in 1991 in Toronto.

- little social participation is expected in the planning process of local well-being, as the limits are decided by science, production (what is produced and how much) and successively ratified by the political level.

The environmental or environmental economic view is a minority school of thought in the field of contemporary economics. It is based on the concept of "efficiency" linked to energy consumption in order to quadruple the productivity of resources (the 50 examples that it proposes are emblematic). The crucial point here is the possibility of decoupling social well-being and development from the concept of resource exploitation (called factor 4), thus restoring "balance to the dynamics of the terrestrial ecosystem". According to this school, the increase in productivity derives from technological evolution linked to the concept of efficiency through:

- corrections to overcome the inconsistencies in the market and in goods, in light of pollution and planning based on the use of technology that promotes efficiency in terms of the use of energy and materials;
- the use of natural capital, according to its regeneration times, and the control of all economic activities, must take into account the sustainability capacity of natural capital.

Some specific practices are:

- Eco-label is an analysis of the environmental impact of a product at all stages of development, from the extraction of raw materials to sale and waste processing. It is only awarded to products that are shown to cause minimal damage to the environment.
- Audit is a procedure for certifying the environmental quality of industrial establishments and their production processes.
- Eco-balances are forms of regulation of codes of conduct and self-responsibility by groups and companies with regard to the environment (some industries present an eco-balance at the end of the year, along with their financial statements).

Ecological economics goes beyond the first two schools, which are based primarily on the concept of "limits". With this school, the term "sustainable" instead takes on a planned and purposeful form. According to the ecological school, "nature and environmental systems have rights that must be respected to avoid the decline of the anthropic system. In this approach, sustainability has a value of reciprocal interaction between human settlement and the natural ecosystem: the future of the human economy is subordinated to the reconstruction of the economy of nature". In ecological economics, the concept of "sustainable" development takes on a precise meaning instead of "compatible" development. Three central problems of the economy are proposed:

- optimal allocation (objective of efficiency, pursued through the instrument of relative prices);
- optimal distribution (objective of equity, pursued through the instruments of income and wealth distribution);

- optimal scale (objective of environmental sustainability, pursued through the political instrument that controls the use of resources, limiting the size of the population and/or per capita consumption).

According to ecological economics school, only the "optimal allocation" of resources can be determined by market prices, but the latter must be associated with a scale and a distribution that pursues objectives of justice and equity. It therefore proposes:

(a) limiting the human scale to a level that, although not optimal, restores the sustainability capacity of terrestrial systems; once this level is reached, growth must be halted by determining a demographic limit and an average standard of living (steady state principle);
(b) increasing technological efficiency more than the flow of materials and energy;
(c) the percentage of use of renewable resources must not exceed the regeneration rate, and waste emissions must not exceed the assimilative capacity of ecosystems;
(d) non-renewable resources may be exploited but at a rate not higher than that which implies the creation of renewable substitutes.

Particular attention is undoubtedly deserved by the territorialist approach.[2] It is promoted by an Italian school composed of architects, sociologists and economists; it deals with self-sustainability and assumes that such a condition can be achieved through a positive reappropriation of the territory by organised groups of civil society in order to confront the process of "deterritorialisation" (territorial disintegration). With this term, it refers to the fact that "the territory, from which man has progressively disengaged himself, considering it a set of negative relationships (environmental, energy, climatic, building, wild location, etc.), due to modernisation, has been treated as a mere technical support for local economic activities and functions organised according to independent principles of relationship with the place, with its environmental and cultural qualities: a quality that derives precisely from its long-standing historical construction. According to this school, throughout a historical period, modernity which culminated with Fordism and mass production, traditional theories of development have considered and used the territory in always reductive terms, denying the value of its intrinsic qualities where:" the producer/consumer has been replaced by the inhabitant, the site by the place, the economic reason by the historical reason. The consequences can be summarized from the following main points:

[2] Defined by the territorialist school for sustainable development. This school was born in the early nineties in Italy by some teachers and researchers of urban planning and sociology who have chosen to coordinate their research activity in the university and CNR environment. These are: A. Magnaghi (University of Florence), G. Ferraresi (Politecnico di Milano), A. Peano (Politecnico di Turin), E. Trevisiol (IUAV), A. Tarozzi (University of Bologna), E. Scandurra (University of Rome La Sapienza), A. Giangrande (University of Rome Tre), D. Borri (University of Bari) and B. Rossi Doria (University of Palermo)".

- the urbanization of industrial peripheries and metropolitan conurbations has devastated place names, towns, cultures, rural landscapes, transforming building areas into zoning and macro functions of the productive and reproductive cycles of large factories;
- mountain and hill sites, important for their size and history (for example, the important civilization of mountain cities), have been contaminated and culturally destructured by abandonment, by decontextualized urban models, and by the location of the peripheral and harmful functions of the productive system;
- some plains, completely cleared to be used by the "green industry", thus transforming the variety of rural landscapes into a mechanical-chemical desert;
- Coastal sites that are mono-functional to the "leisure" "pleasure/fun" of the mass producer/consumer, with the continued construction of the coastline and the devaluation of the mountain landscapes of the internal areas.

This approach differs from all others (including that of ecological economics) in that it pays greater attention to the local dimension and considers environmental sustainability to be inseparable from the cultural, social, political and economic aspects of the territory.

In this context, the territory is conceived as a historical product of long-term co-evolutionary processes, between urbanization and environment, between nature and culture, through continuous and stratified cycles of civilization. These processes generate a territorial space (sets of geographic sites) characterized by a "temporal depth", in terms of local identity, distinctive characteristics, local specificities: that is, highly complex living systems. To counteract the process of de-territorialization, the activation of "virtuous relationships" between the components of the territory is proposed: anthropic environment, built environment and natural environment. The concept of sustainable development therefore depends on the following factors:

- quality and replicability of natural resources (natural sustainability);
- complex and interrelated systems, including non-hierarchical organization of urban systems (urban sustainability);
- coherence of productive systems with the valorization of territorial heritage and with the development of the local business world (economic sustainability);
- growth of forms of self-government of local societies (socio-cultural and political sustainability).

The term "local" reveals the valorization of territorial resources and the identity of a place; the term self-sustaining, on the other hand, indicates the importance of a search for settlement, economic, political-social rules that produce local homeostasis and long-term balance between the natural environment, the built environment, and the anthropic environment Magnaghi (1990).

4.2.2 Social Crisis

By social crisis instead we must refer to the decoupling of wealth from well-being, from citizenship rights. In fact, Daly and Cobb[3] study (1989) shows that, from the second half of 1970s in the United States, the growth of monetary wealth measured through the Gross Domestic Product no longer corresponded to real welfare, ISEW, (which included services and distribution of wealth), but with new poverty of rights and conflicts. A situation that is also observed both in problems with the environment (externalities[4]) and with the phenomenon called jobless-growth.[5] Even unemployment in the private sector is at serious risk, according to data from the World Bank Report, from 1975 to 2012 there was a drop in the employment rate in businesses, of 75% in advanced countries and 59% in developing countries (The World Bank Annual Report, 2019).

As Hobsbawn states (2010, starting from the industrial revolution, capital and labor take two different paths. The affirmation of mercantile enterprises as producers of monetary wealth, mega profits, has led the economy to distance itself from society, putting its rights, even human, at risk. In the 1990s, in fact, new socio-economic practices were born also in terms of Bourdieu (1980), such as social economy enterprises, enterprises that try to combine the economy with a vision of life that had remained outside also due to the process of economic positivization. From an analysis by Jeremy Rifkin (1995) of those years, it is highlighted how the socially oriented economy, or third sector, has a growth rate twice as high as the traditional economy, although forms of social economy had existed in previous centuries (Vuotto, 2006). The social economy also defined as non-profit, due to the characteristic of the non-distribution constraint (Hansmann, 1980, 1986), arises from the need to reinsert into the economy a vision of life that has now been lost by the economic positivism that has reduced everything to mathematical calculations aimed at maximizing monetary interests (Vigliarolo, 2019).

---

[3] The study by Daly and y Cobb (1989) showed that, from the second half of the 1970s, in the United States, for example, the growth of monetary wealth measured through the Gross Domestic Product did not correspond to real welfare (which included services and distribution of wealth), but to new poverties that affected the main rights of the country's citizens. In other words, the loss of the relationship between "traditional" economic growth and the real well-being of a society is one of the main problems of this times.

[4] In economics, it means the positive or negative impact that economic actions have directly or indirectly at a social and environmental level is called externality. In this case, it is enough to just mention global warming due to the emissions of harmful gases (mainly CO_2) caused by the world's industrial systems.

[5] With this term, we understand the phenomenon of economic growth without employment growth, whose first signs go back to the crisis of 1929 in the United States of America but which has strengthened since the first decade of 2000. Cf. Toossi Mitra, (1 December 2013).

4.2.3 A Phenomenological Approach to Economics and the Demand for Socio-Environmental Rights

In this context, some authors propose a phenomenological, approach to interpreting the economy also in terms of ideas that aim to build a vision of life. For those who write, economic phenomenology can be defined as the branch of phenomenology that aims to study the economy in the formation of its main ideas, which contribute to the creation of the structure and vision of the life of a country, a society, starting from the pure consciousness that opens to the world, through the relationship between noesis and noema. In this direction, economic phenomenology studies the coherence of economic systems with respect to these ideas considered as priorities in human consciousness that can become social, and how they can represent the structures of an ontological reason that allows the implementation of what we call social being in the world (Vigliarolo, 2019).

For ontological reasons, in these pages, it is understood as the set of those characteristics that define human beings in their structural or constitutive dimension, thus conceived in consciousness. In other words, it refers to a relational dimension of all parts of the context, referring to the territorialist approach of sustainable development. For example, without a healthy relationship with the environment, which depends on human behavior, human life is also put at risk. For these reasons, we speak of a demand for socio-environmental rights of a people, conceived in social consciousness.

The term "demand for rights" therefore defines the set of process, materials, goods and services that can be reduced to primary ideas[6] that the socio-institutional system considers important and that the economic system carries forward. The process of defining a demand for socio-environmental rights, therefore, is relational by definition and implies the constitution and "awareness" of the human, social and environmental priorities that must guide the economy. The demand for rights responds to the identities of all the parties in a context, and therefore cannot be defined by individual interests.

4.3 Some Defining Aspects on Circular Economy

In this context, we try to propose the circular economy, i.e. as a process of affirmation of the ontological reason of peoples. Before continuing, which circular economy are we referring to in these pages? A simple definition of it says that "A circular economy is a clean production system designed to minimize the use of resources in production systems and to reduce unwanted externalities from production

[6] For example, the idea of eating, moving, dressing, etc. These are ideas defined as primary or general. The way in which one eats, moves, dresses, are defined as particular or secondary (Vigliarolo, 2019).

processes, including waste, emissions and energy losses". As we can see, attention is placed on the fact that the production process must be made with materials that can be recycled. Continuing, always according to Camilleri, this economy proposes a completely different approach to the economy of growth. "Such a sustainable approach differs from linear economic approaches that are based on the premise of "take-make-consume-dispose" actions (Camilleri, 2019, 2023), which assumes that resources are abundant, available and cheap to dispose of".

The term "circular economy" was first used in Western literature in 1980 (Pearce & Turner, 1990) to describe a closed system of interactions between economy and environment. The circular economy is part of the study of feedback of non-linear systems, considered as living systems.

The model of the 3Rs (reduce, recycle, reuse) can be considered the idea underlying the circular economy which is diametrically opposed to the linear economy based on the linear flow of materials, return-product-residues, which need to be transformed in a circular flow of recycled product-return. Indeed, the theme of residues is a central issue of the circular economy. It is expected that over the next 40 years the overall consumption of materials such as biomass, fossil fuels, metals and minerals will double, and in parallel the annual production of waste will increase by 70% by 2050.[7]

According to a definition by the Ellen MacArthur Foundation, the circular economy «is a generic term to define an economy designed to be able to regenerate itself. In a circular economy, there are two types of material flows: biological ones, capable of being reintegrated into the biosphere, and technical ones, intended to be revalued without entering the biosphere». The circular economy is therefore an economic system planned to reuse materials in subsequent production cycles, reducing waste to a minimum. According to the Ellen MacArthur Foundation there are 5 dimensions.

4.3.1 Eco-Design

Designing products by immediately thinking about their use at the end of their life, therefore with characteristics that will allow them to be dismantled or renovated.

4.3.2 Modularity and Versatility

Giving priority to the modularity, versatility and adaptability of the product so that its use can adapt to changes in external conditions.

[7] World Bank (2018), What a Waste 2:0: A Global Snapshot of Solid Waste Management to 2050.

4.3.3 *Renewable Energy*

Relying on energy produced from renewable sources, encouraging the rapid abandonment of the energy model based on fossil fuels.

4.3.4 *Ecosystem Approach*

Thinking holistically, paying attention to the entire system and considering the cause-effect relationships between the different components.

4.3.5 *Material Recovery*

Promoting the replacement of virgin raw materials with secondary raw materials from recovery chains that preserve their qualities.

The circular economy thus conceived, according to the architecture of these pages, therefore allows us to promote what we define as the reciprocity of relations between the anthropic community and the natural context, absolutely necessary to promote the relational and constitutive rights of a society. It allows us to move away from acting only with personal interests, which as we have seen lead to serious environmental and social crises. For these reasons, it is important to understand how the dichotomy personal interests vs. reciprocity arises.

4.4 Reciprocity vs. Personal Interests

4.4.1 *Before Smith: The Physiocrats and the First Cyclical Model*

The concept of circularity of the economy dates back to the Physiocrats, from whom Smith also took the concept of *laissez-faire*.[8] With the physiocratic school, founded by the physician François Quesnay with his work Tableau économique of 1758, the first cyclical economic model was theorised, which laid the foundations for the concept of an autarkic economic system. In fact, it was the first school to conceive economic activity as a continuous flow of income circulating from one class to another, which could be represented by a synthetic table, the "Economic Table" (precisely, the Tableau économique). This cycle was based on the existence of a

[8] *Cfr.* Miglio (2001).

natural law and on the defence of property and free trade (laissez-faire or non-public intervention in the economy), which allowed income to pass from one class to another.

The physiocratic contribution is important because it establishes the foundations for the concept of macroeconomic equilibrium, where the means of production are invested, and then, the products are put back into circulation to be invested again. It is a circular model that leads Marx to say that the true founder of classical economics is Quesnay and not Smith, who will take the idea of the free market from the physiocrats (Vigliarolo, 2019).

A little later, a contemporary of Adam Smith, Genovesi, in the eighteenth century founded what is called civil economy, which is based on the concept of reciprocity that Bruni and Zamagni take up again in these years (2004). A civic (or citizen) economy does not aim at maximizing well-being defined in an abstract and utilitarian way, but at "public happiness", (Bruni & Zamagni S., 2004). It is the result of the intertwining of the economy with social relations, as Polanyi also argued (1947). The understanding of economic action includes value-based and reciprocal motivations, in addition to those of an interested or instrumental nature. It is a criticism of the neoclassical paradigm that limits the motivation of economic action to the latter type. Starting from the hypothetical-deductive method (in force substantially since Ricardo and Stuart Mill), the economy is conceived starting from the concrete life of people and communities. These authors argue that for the economy to function harmoniously, not only the exchange of equivalents (based on the contract) and coercive redistribution relationships (taxes and transfers) are necessary, but also the gift and reciprocity. Furthermore, the functioning of the market itself requires the "principle of reciprocity", otherwise the market itself risks disappearing due to lack of trust (Octavio Groppa, 2016).[9]

4.4.2 Smith's Arrival

The cultural battle between Genovesi and Smith is undoubtedly won by the latter due to a fundamental historical fact: the arrival of the industrial revolution.

With Smith, the concept of reciprocity and circularity is definitively lost. The individual becomes the centre of the economy with his individual interests. The context is conceived as functional to individual interests and the exploitation of resources is conceived as endless, due to the concept of permanent growth of economic systems. In this context, the tension between individual good and common good is resolved in Smith with the famous invisible hand that could never be proven; but for utilitarians it can be measured as the sum of individual pleasures/goods. We could say that the figurative representation of this model is linear, like the concept

[9] Octavio Groppa, 2016. Economía civil, bienes comunes y reciprocidad en Economía, Centro de Estudios en Economía y Cultura de la Universidad Católica Argentina.

of progress itself that was developed from classical and neoclassical theories, with marginalist theories. Finally, the cultural basis that permeates the Smithian school is undoubtedly economic positivism with which the same economic systems are cleansed of any transcendental or ideal dimension. That is, the subjectivation of systems as a human decision. It is for these reasons that in 1891 Keynes Father Keynes (1999, ed. 1891) defined the difference between positive economics and normative economics.

4.4.3 Keynes and his Limits of Generic Aggregate Demand

In this scenario, John Maynard Keynes, the son, tries to put a limit on this model. He introduces a collective organism, like the State, to the individualistic Smithian model and through this, completely reverses the microeconomic foundations, proposing the aggregate demand for consumption of economic systems. With this, in fact, macroeconomics is founded. What Keynes never delved into was undoubtedly the attention to production. In fact, many authors claim that there is no true productive theory in Keynes. Aggregate demand contemplates any type of goods and services, such as the production of weapons or goods with high negative externalities, which seriously affect community life. Both, for example, are economies of death, one thinks of wars and the high externalities generated by toxic waste. Just to mention a few examples. Over the years, the economy becomes the enemy of citizens. It no longer responds to a human, social and environmental identity with which it has to measure itself. But to the accumulation of capital without end.

4.4.4 Economic Financialization

With the industrial revolution, a new era undoubtedly begins, that of financial capital. If before it was linked to production, since 1970 financial capital has a life of its own. In fact, from the 1970s onwards, with Keynesianism in crisis, neoliberalism returned, this time under the financial semblance. With the fall of the active production rate in the USA in the 1970s, the golden years came to an end. Nixon untied the dollar from the gold standard to allow economic groups in his country to make huge profits in the financial markets Arrighi (1996, 2007). In this way, financial transactions multiplied (Álvarez Peralta, 2012) and today the relationship between the growth of the world GDP and financial values is 10 to 30. The economy has lost its social orientation and what we called before the relationship with ontological, social and environmental reason. How can we recover all this and generate an ontological reason between a micro-economy that responds to personal interests and a macroeconomy that has been reduced to a generic consumer demand? Is it possible to interpret the economy in terms of ideas and orient it from a social conscience towards a socio-environmental reason?

4.4.5 *New Models and Challenges for Economic Science: A Generative Economy of Ideas*

As we have seen before, economic phenomenology deals with ideas that are built on the basis of economic systems. We can affirm that there is a set of general ideas that are part of social knowledge and considered necessary for all societies so that primary socialization can take place, such as feeding, healing, educating, training, moving, etc. What changes are the particular ideas, or the ways in which they are implemented, as we see before. The nucleus of general ideas present in all societies can be considered universal. That is, all societies develop the idea of eating, dressing, educating, moving, etc.. Furthermore, it is it possible to set that all societies can generate their own life ideas that allow to create a knowledge and orient the economy.

4.4.6 *Mesoeconomy*

This process could take place in a meso-economic dimension, a space between micro (individual interest) and macro (generic demand of good and services). Since the process of social knowledge formation is intersubjective (Schutz, 2003), the meso-economy space is made up of organizations, companies, civil society, and the State, ... where the cultural elements that orient the economy towards the construction of community life and identity are built and incorporate in economy. It means, these elements are defined and socialized in the form of goods, services, norms, rules, ideals, codes, practices, etc. shared by the local or global "socio-economic pacts", which allow the economy to recover again its normative/ontological dimension and promoting the rights of all community (Vigliarolo, 2017).

4.5 Conclusions

The importance of reducing the economy to ideas is fundamental to being able to make choices and make actions intelligible in order to orient and direct the economic system. For example, a company is always born, lives and develops in a context. It also feeds on relationships with other actors and with the surrounding environment. It is subject to national and international legal norms within a political, social and environmental order that defines it, surpasses it and contains it. It is made up of workers with their own human and social history who live their own biographical dimension shaped by other individuals, all inserted in a social and environmental context. In this context, the company is called to understand and grasp the identity of the entire context. That is, it can seek to maximize profits or promote an ontological function.

In other words, the choice to become a circular company depends on the ability to become aware of all the dimensions that surround it - environmental, social and economic - and to fuel a demand for socio-environmental rights and not a demand for consumption. All this implies that the market itself must be a subset of society (embodied) and can never be the main driver that defines the ontological elements of human life. Circular economy companies contribute to the promotion of a demand for rights, while the market economy responds to satisfy a demand for consumption.

All this can be oriented through a meso-economic approach. In this way, the economy can generate a circularity of materials, but also and mainly of ideas that are the basis of the promotion of community rights that are inscribed in the conscience of peoples and allow their permanent change.

Norms, social meanings, etc. are defined permanently. They are absolutely necessary to define an institutionality necessary to promote a demand for socio-environmental rights for progressive degrees of social and environmental awareness, which cannot be left in the hands of individual wills.

In this way, the strength of a circular economy company depends on its ability to read and create, reproduce and generate reciprocal relationships, which are interpreted in terms of ideas and allow it to become aware of its ability to produce benefits in harmony with the natural dimension that surrounds it.

References

Álvarez Peralta, N. (2012). Financiarización, acumulación de capital y ajuste salarial. In *XIII Jornadas de economía critica*, jornadas llevadas a cabo en Sevilla, España.

Arrighi, G. (1996). *Il lungo XX secolo: denaro, potere e le origini del nostro tempo*. Il Saggiatore.

Arrighi, G. (2007). *Adam Smith en Pekín. Orígenes y fundamentos del siglo XXI*. Akal.

Bourdieu, P. (1980). Le capital social. Notes provisoires. Actes de la Recherche. *Sciences Sociales*, N° 31.

Braudel, F. (Ed.). (1986). *I tempi della storia economica, società, civiltà*. Dédalo.

Braudel, F. (Ed.). (2002) [1985]. *La dinámica del capitalismo*. Fondo de Cultura Económica.

Bruni, L., & Zamagni, S. (Eds.). (2004). *L'economia civile*. Il Mulino.

Camilleri, M. A. (2019). The circular economy's closed loop and product service systems for sustainable development: A review and appraisal. *Sustainable Development, 27*(3), 530–536. *(PDF) Circular Economy*. Available from: https://www.researchgate.net/publication/375809545_Circular_Economy#fullTextFileContent. Accessed 26 Oct 2024.

Camilleri, M. A. (2023). Circular economy. In *Encyclopedia of sustainable management*. https://doi.org/10.1007/978-3-031-25984-5_399

Daly, H., & Cobb, J. (1989). *For the common good*. Beacon Press.

David, T. (Ed.). (2005). *Economía e Cultura*. Il Mulino.

Donati, P. (Ed.). (1998). *Teoria relazionale della società*. Franco Angeli.

Durkheim, E. (Ed.). (1976). *Educación como socialización*. Sigueme.

European Commission. Towards a circular economy. https://ec.europa.eu/commission/priorities/jobs-growth-and-investment/towards-circular-economy_en

Hansmann, H. (1980). The role of non-profit enterprise. *Yale Law Journal, 89*, 835–898.

Hansmann, H. (1986). The role of nonprofit enterprise. In S. Rose-Ackerman (Ed.), *The economics of nonprofit institutions: Studies in structure and policy*. Oxford University Press.

Hilferding, R. (Ed.). (1973). *El capital financiero*. Ediciones el Caballito.
Hobsbawm, E. (Ed.). (2010 [1975]). *La era del Capital 1848–1875*. Crítica.
Hounie, A. (Ed.). (Comp) (2010). *Sobre la idea de comunismo*. Paidós.
Keynes, J. N. (1999 [1891]). *The scope and method of political economy*. Batoche Books.
MacArthur, E. Circular economy overview—The concept of circular economy, Isle of Wight, United Kingdom. https://www.ellenmacarthurfoundation.org/circular-economy/overview/concept
Magnaghi, A. (1990). *Il territorio dell'abitare: lo sviluppo locale come alternativa strategica*. Franco Angeli.
Miglio, B. (2001). *I fisiocratici*. Laterza.
Octavio Groppa, 2016. Economía civil, bienes comunes y reciprocidad en economía, centro de estudios en economía y cultura de la universidad católica argentina
Palley, T. (2007). Financialization: What it is and why it matters. *Political Economy Research Institute*, N° 153.
Pearce, D.W. and R.K. Turner (1990), Economics of Natural Resources and the Environment, Hemel Hempstead: Harvester Wheatsheaf.
Perna, T. (1998). *Fair trade. La sfida etica al mercato mondiale*. Bollati Boringhieri.
Perozzi, G. (2015). *La finanziarizzazione dell'economia statunitense: origini, logiche e conseguenze*. Travail de Bachelor, Université de Fribourg - Faculté des Sciences Économiques et Sociales.
Piketty, T. (2013). *Le Capital au XXIe siècle*. Éditions du Seuil.
Polanyi, K. (2000). *La grande trasformazione*. Einaudi.
Rifkin, J. (1995, Ed. 1997). La fine del lavoro, il declino della forza lavoro globale e l'avvento dell'era post-mercato, Milán: Baldini & Castoldi.
Schütz, A. (a cura di). (2003). Problemi della realtà sociale. Scritti I. Buenos Aires: Amorrortu.
Stiglitz, J. (2010). *Caída libre: El libre mercado y el hundimiento de la economía mundial*. Taurus.
Stiglitz, J. (2012). *El precio de la desigualdad. El 1% de población tiene lo que el 99% necesita*. Taurus.
The World Bank Annual Report 2019: Ending Poverty, Investing in Opportunity.
Vigliarolo, F. (2017). *Lupi e Agnelli, società capitaliste e vie d'uscita*. Jaca Book.
Vigliarolo, F. (2019). *La economia es un fenomeno social. Principios de fenomenología económica*. EUDEBA.
Vuotto M. (compiladora). (Ed.). (2006). *Economía social. Precisiones conceptuales y algunas experiencias históricas*. Editorial Altamira.
World Bank. (2018). What a Waste 2:0: A global snapshot of solid waste management to 2050.
Zohc, A. (Ed.). (2006). *Lo spirito del danaro e la liquidazione del mondo*. Milán Jaca Book.

Francesco Vigliarolo is full time prof. of Regional Economics at Catholic University of La Plata and social economist. PhD. in Social Sciences, specialization in economic sociology, post doctorate in Economics and Master in sustainable development. He is also an Associate Investigator of the CESOT (Centro de Study of Sociology of Work) of the University of Buenos Aires. At the Universidad Nacional de La Plata he is Chairolder of UNESCO chair in "Economic Systems and Human Rights" he founded with almost 20 Universities all over the world. From 2008 to 2015 he worked at the University of San Martin where he promoted and designed the first Master course financed completely with European funds in Argentina in "Democratization and Human Rights". Before moving from Italy to Argentina, he worked at the University of Ferrara where he was professor of Social Economics and Methodology and Technique of Social Investigation and coordinator of some Masters, including "Social economics and non-profit organizations" and "Social economics and international cooperation". Always in Italy, he has also been Director of CIPSI, Federation of 45 NGOs present in more than 60 countries around the world. With his investigations, he is interested in local development processes such as the construction of territorial identities that imply the affirmation of rights (ontological reason) through the strength of associative and democratic forms of production. In this direction, he works to what is called an "economic phenomenology" which aims to observe the economy starting from the relationship between subject and materiality, which presupposes "functions, ideas, concepts" and the affirmation of the "relational identity of a democratic society", which defines an ontological reason as opposed to the utilitarian one. He published various books and scientific articles from which he explains concepts such as "social ethical capital", "economic socialization and democracy", "demand for the rights of the people", "meso-economics approach" and "the principles of economic phenomenology." As his latest books we mention "The economy is a social phenomenon. Principles of economics phenomenology" (2019) published with Eudeba, "Wolves and lambs. Capitalist societies and ways out" (2017) published with Jaca Book. Personal blog: http://fvigliarololupieagnelli.blogspot.com/; Website UNESCO CHAIR https://unlp.edu.ar/unesco/

Chapter 5
Balancing Environmental, Social, and Economic Dimensions: Marginalization of Katuic's Ethnic Minorities in Vietnam and the Adaptive Circular Economy

Dinh Hong Hai and Le Anh Tuan

Abstract Establishing a circular economy is crucial for sustainable development and has gained global government attention. The circular economy in Vietnam is being implemented in various sectors, particularly high-tech development areas. However, remote areas require a different approach due to their need for more scientific and technological frameworks. In Central Vietnam, where the Katuic language group, comprising the Katu, Bru-Van Kieu, and Taoih ethnic minorities, resides, a "natural circular economy" production method has been prevalent for a long time. This method is comprised of traditional "slash and burn" agriculture (shifting cultivation), but the impact of industrialization, modernization, and urbanization in Vietnam after Doi Moi has led to the marginalization of the Katuic people. Currently, the marginalization of the Katuic people is increasing due to the profound impact of a lack of resources on the development process. Moreover, they are influenced by the rapid development of national renewable energy policy, such as wind turbines and hydroelectric dams surrounding their habitat. Many communities must move to the new government resettlements without forest, river, or rice fields. A balanced solution addressing environmental, social, economic, and cultural aspects is essential to enhance their quality of life and establish a sustainable future. This research examines the impact of the development process in Vietnam and the marginalization of the Katuic group and proposes a specific development solution called the "Adaptive Circular Economy." The latter approach, which addresses natural and human factors, is situated in household micro-economic cells.

Keywords Balancing dimensions · Katuic marginalization · Adaptive Circular Economy (ACE) · Household Micro-Economic Cells (HMEC)

D. H. Hai (✉) · L. A. Tuan
National University, Hanoi, Vietnam
e-mail: haidinh@vnu.edu.vn

S. Patnaik et al. (eds.), *Circular Economy and Sustainable Value Creation through Eco-Innovation*, Circular Economy and Sustainability,
https://doi.org/10.1007/978-3-031-88516-7_5

5.1 Introduction

The circular economy is now vigorously promoted as a sustainable development model to address environmental challenges and resource depletion. Instead of using the traditional linear economic model ("take-create-waste"), the circular economy focuses on extending product life cycles, optimizing resource use, minimizing waste, and recycling. This helps conserve natural resources and reduce waste in the environment. The circular economy model is the future of sustainable economics, not only helping to protect the environment but also opening up new business opportunities through the creation of products and services with longer-term value.

According to the European Parliament, "the circular economy is a model of production and consumption, which involves sharing, leasing, reusing, repairing, refurbishing and recycling existing materials and products as long as possible. In this way, the life cycle of products is extended. In practice, it implies reducing waste to a minimum. When a product reaches the end of its life, its materials are kept within the economy wherever possible, thanks to recycling. These can be productively used again and again, thereby creating further value. This is a departure from the traditional, linear economic model, which is based on a take-make-consume-throw-away pattern. This model relies on large quantities of cheap, easily accessible materials and energy" (European Parliament, 2023).

At the macro level, the Ellen Macarthur Foundation has identified three main principles of a circular economy, which include 4Rs: Reduce, Refurbish, Repair, Reuse with the main goal of extending product life cycles, minimizing and eventually eliminating waste: (1) Reducing and eliminating waste and pollution; (2) Extending the shelf life of products and materials; (3) Regenerating natural systems. In general, a circular economy aims to extend the useful life of products, equipment, and infrastructure to increase the efficiency of these resources. All the "waste" of one consumer production process is considered raw material for other consumer production processes, whether it is a by-product, a resource recovered from another industrial process, or a resource returned to the natural environment, such as composting organic waste. In Vietnam, the circular economy is becoming a focus of sustainable development strategies. Vietnam has introduced initiatives and policies to promote this model, with industries such as textiles, agriculture, and food processing beginning to apply circular principles to minimize waste and use resources efficiently. However, a "natural circular economy"[1] in minority or tribal communities like Katuic people[2] needs to be based on indigenous knowledge, social resources, human resources, and community institutions embedded in customary law and local conventions.

[1] In the traditional economic model of the Katuic people, the cycle of waste is recycled back into the natural environment, which is the "natural circular economy."

[2] The Katuic people in Katuic language group includes the Katu, Bru-Van Kieu, and Taoih, mainly living in Central Vietnam and neighbour provinces of Laos.

The Katuic people have used the traditional agricultural method of "slash-burn" (or multi-cropping, rotation, intercropping) for thousands of years. This form of production, using Indigenous knowledge, comprises an effective economic cycle, creating a balance between the natural environment, humans, and other living species. Farming is wholly waste-free and pollution-free because raw materials are used in rotation. The level of regeneration is nearly complete, almost reaching the natural level. In reality, the traditional knowledge, biodiversity, and value from fallow have been applied in many countries around the world as a key facilitating factor for maintaining sustainable subsistence and food security with "the rich traditional knowledge of the natural environment, and land use and management practices that have been developed over generations and are well adapted to the environment. The strength of traditional land-use systems, shifting cultivation in particular, lies in the diversity of the locally adapted practices and crops grown. Many local varieties of domesticated plant species adapted to local environmental and climatic conditions are grown using traditional shifting cultivation. The diversity of local domesticated plant varieties and also the biodiversity of fallow forests represent a rich genetic pool of useful plant species. This can play a crucial role in adapting local livelihoods to the human-induced changes that our climate is currently undergoing" (Erni, 2015, 22).

Before industrialization, modernization, and urbanization (IMU) took place in Vietnam, the forest cover rate was relatively high. "The total forest area of the whole country was recorded as 14.3 million hectares (in 1945), however, in 1995, the natural forest was encroached, transformed the purpose of use and excessive exploitation, so the area was only 8.25 million hectares" (Huy Hoàng, 2017). After Doi-moi, the development process required increasingly large land areas to build industrial parks and power plants. When this process took place in the residential area of the Katuic group over the past three decades, it clearly showed that deforestation was due to IMU, not shifting cultivation. In fact, after more than three decades of industrialization, modernization, and urbanization in Vietnam, the above-mentioned traditional balance factors have been completely broken due to changes in the residential environment, production conditions, social relations, and cultural and religious changes. These changes were caused mainly by migration into the area', and the Katuic residents became marginalized entities.

Due to this marginalization, most Katuic people are left behind, finding it increasingly difficult to adapt to the modern economic development process. These people themselves have a passive life with an inferiority complex of ethnic minorities, so it is tough for them to access national resources in the context of the industrialization and modernization of Vietnam. To help them escape this situation, many programs and projects of the Vietnamese State and international organizations such as the World Bank (WB), United Nations Development Programme (UNDP), Asian Development Bank (ADB), and International Fund for Agricultural Development (IFAD) have been implemented. However, it seems that the value chain of these programs and projects is not enough to help them improve their economic life because most people have a low level of education. Therefore, our bottom-up approach is to solve this bottleneck directly.

The question that needs to be asked is how to balance the environmental, social, economic, and cultural aspects to transform the subsistence of the Katu people to adapt successfully to the development process. To answer the above question, this study will investigate the economic, environmental, and social status of the Katuic language group, including Katu, Bru-Van Kieu, and Taoih (referred to as Katuic in this study) in the context of Vietnam's economic boom. By analyzing the marginalization of this minority group, we propose a balanced solution: an adaptive circular economy plan placed in the microeconomic entity of households based on indigenous knowledge. Based on this model, the research will pilot Family Micro-Economic Cells in some Katu, Bru-Van Kieu, and Taoih households in Nam Đông District, Quảng Nam Province, A Lưới District, Thừa Thiên – Huế Province, and Hướng Hóa District, Quảng Trị Province (Table 5.1).

In 2022, we surveyed the Katuic residents in Nam Đông District, Quảng Nam Province, A Lưới District, Thừa Thiên – Huế Province, and Hướng Hóa District, Quảng Trị Province (Đinh Hồng Hải, 2022). This pilot project was in some households of the Katu, Bru-Van Kieu, and Taoih people (see more detail in Sect. 5.4). Research results (Tables 5.2, 5.3, 5.4 and 5.5) from 2022 are the foundation for conducting this research. Accordingly, to develop models of transformation from self-sufficiency to modern agriculture according to the policy of the Vietnamese government, it is impossible to apply the model of people in the plains.[3] The

[3] "Improving the mountainous region to catch up with the plains region" (đưa miền núi tiến kịp miền xuôi) is the slogan based on the Resolution of the 15th Central Conference of the Communist Party of Vietnam (January 1959), which set out the task: "Our Party and State need to have a com-

Table 5.1 The impact of industrialization, modernization, and urbanization in Vietnam after Doi moi

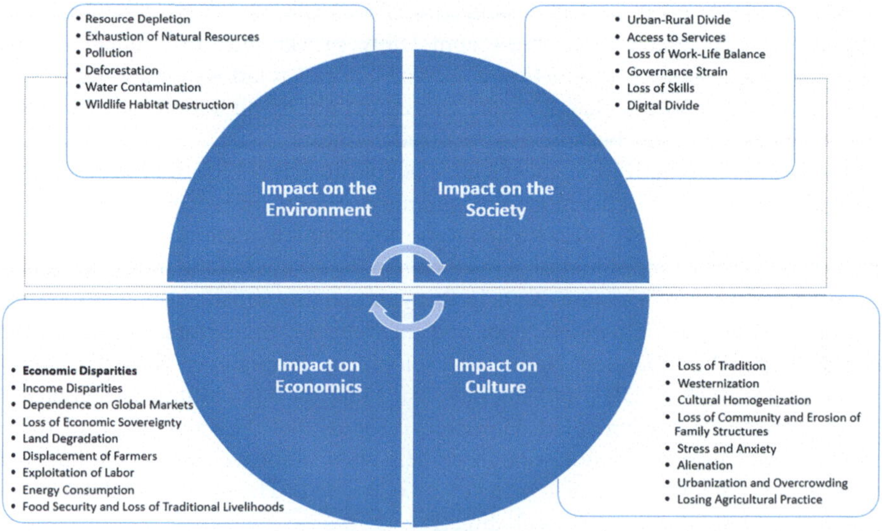

Table 5.2 The people's anxiety about the electric project in CEZ (including A Luoi District, Nam Dong District, and Huong Hoa District). (Source: Đinh Hồng Hải, 2022, 17)

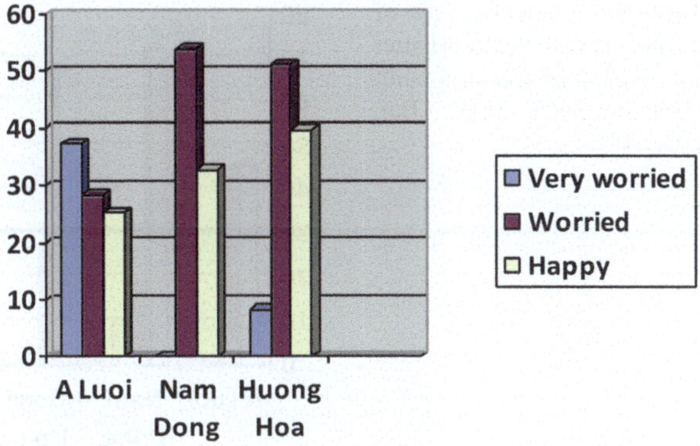

Table 5.3 The regret of the people about the loss of rivers, streams, forest, land, and rice fields because of the electric project in CEZ. (Source: Đinh Hồng Hải, 2022, 17)

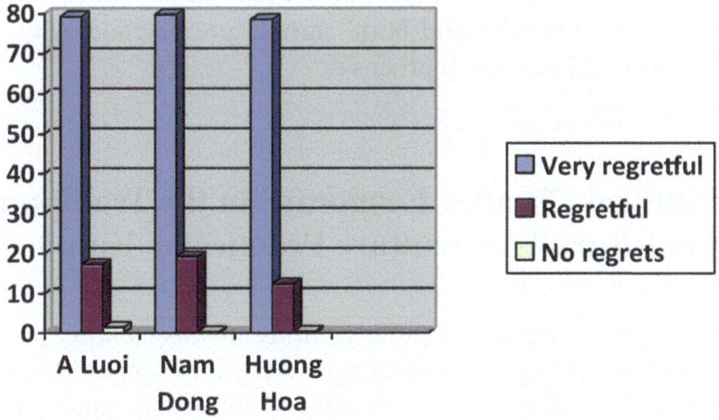

Table 5.4 The care of the people with the forest after the electric project was built in CEZ. (Source: Đinh Hồng Hải, 2022, 18)

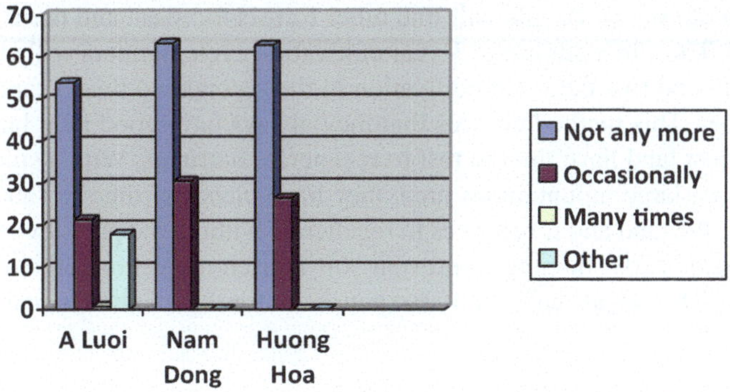

Table 5.5 The connection of the people with the forest after the electric project was built. (Source: Đinh Hồng Hải, 2022, 18)

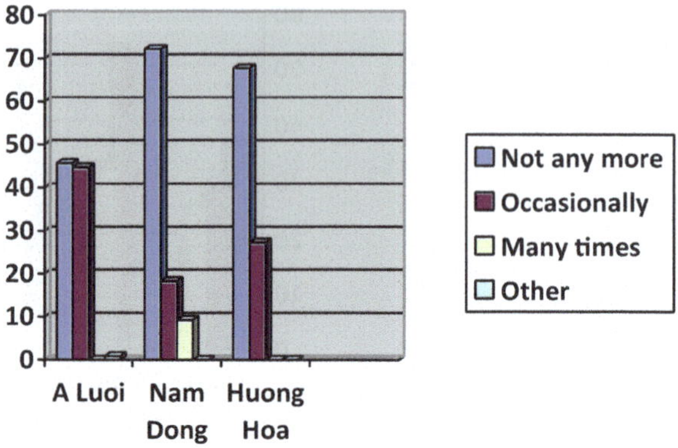

development model of minority groups such as Katuics needs to have its characteristics. Therefore, the top priority must be balancing the environmental, economic, social, and cultural status factors. To balance the above factors, the "natural circular economy" in the traditional "slat-and-burn" agriculture practice by the Katuic people was chosen during this research process.

5.2 The "Natural Circular Economy" in the Traditional "Flat-and-Burn" Agriculture Practice by Katuic People

"Slash and burn" agriculture is a popular farming method among people living in mountainous and tropical regions, typically in Africa, South and Central America, and Southeast Asia, including Vietnam (Peters & Neuenschwander, 1988; Đỗ Đinh Sâm, 1994; Kleinam et al., 1995; Vosti & Witcover, 1996; Stefan & Lindsey, 2013). This is a manual agricultural method, using knives to cut down trees on the hillside, then drying and burning to get soil to plant trees. The coal ash is a natural fertilizer for crops. After 2–3 years of cultivation, the field will be abandoned to allow the soil to regenerate nutrients. People will find other forests to "slash and burn" to replace the old field. Thus, in a minimum 3-year cultivation cycle, farmers will always have one new field and two fields for cultivation in the second and third years, exchanging each other. This method ensures that they always have good field land but also helps the forest land have time to rest to regenerate nutrients. With a small population living in a large mountainous area, they have plenty of time to "abandon" the old fields for the land and forest trees to regenerate without the pressure of returning to the old fields early, helping to prevent soil nutrient depletion, erosion, and soil degradation. This is an easy and inexpensive farming method suitable for the

prehensive plan and long-term economic and cultural development in mountainous areas progress in the plains, helping ethnic groups to promote their revolutionary spirit and great ability" (Communist Party of Vietnam Resolution 15th, January 1959).

economy of households and small-scale communities. This farming method ensures ecological sustainability based on the balance between crop yield and soil fertility.[4]

For sequence, "shifting cultivation" in production and "nomadic" in daily life have become adaptive methods for the Katuic group to survive in the mountainous environment for a long history and what is still current. The "natural circular economy" was still strongly preserved in the Katuic group until before Doi Moi, as our explanation in recent research (Nguyễn Hữu Thông, Lê Anh Tuấn et al., 2005, 158). The presence of a "natural circular economy" is formed based on the specific natural and social conditions of the central highlands of Vietnam in Quảng Nam, Quảng Trị, and Thừa Thiên Huế provinces. It is an independent economy with a self-sufficient nature within the village, operated by shifting cultivation and exploiting mountain and forest products. The organic relationship between "nomadic villages" and "shifting fields" contributes to shaping the typical socio-economic background of the Katuic group, which is "shifting cultivation."

The "natural circular economy" ensures the rights and obligations of community members, with the means of production being mountains, forests, and land, and slash-and-burn farming being both a source of livelihood and a basis for establishing residential territory to consolidate village community members. "The material basis for the unique social institution in ethnic minority and mountainous areas is the village community's collective ownership of land and forests" (Nguyên Ngọc, 2002, 5–12). The people of the Katuic group rely on the "nương lịch" or seasonal calendar, which shows a high level of synthesis. It reflects the laws of nature, plants, animals, birds, and animals and applies to household and community economic units. The seasonal calendar (nương lịch) is formed by summarizing experiences of activities that have been repeated for thousands of years into an optimal and most effective "operation plan".[5] The economic cycle is expressed through the relationship and transformation between the type of production activities (cultivation and animal husbandry) and natural exploitation activities (hunting and gathering) in a year with four seasons. This reflects the natural cyclical nature based on traditional indigenous knowledge when their economic cycle is based on the natural cycle. This has helped the people in the Katuic group proactively conduct their productive activities, creating little change or fluctuation over the past thousands of years.[6]

---

[4] See more detail in: Levin S.A. (ed.) *Encyclopedia of Biodiversity*, second edition, Volume 6, pp. 551–562. Waltham, MA: Academic Press, pp. 551–561.

[5] For example, in January, when the weather is cold and rainy, the appropriate work is to rest, organize festivals, get married or go hunting; in February and March, when the weather gets warmer, they start to find fields and clear the land for pruning; April is the latest time to sow rice because in May and June there are often thunderstorms and strong sunshine, suitable for collecting honey, firewood, weaving, taking care of rice grass, etc. In August, they start harvesting because, after September, the weather is rainy and cold, suitable for collective hunting trips (Lê Anh Tuấn 2004; Nguyễn Hữu Thông et al. 2005, 157).

[6] The "natural circular economy" cycle consists of "clearing - cutting - burning - planting - harvesting," repeated yearly. The cycle begins with the men's work of finding fields and ends with the women's work of harvesting. This is how the Katuic group prevents soil erosion during the rainy season and stores water through layers of rotten leaves for crops. On sloping fields, intercropping

For the Katuic people, the "slash-and-burn" farming method represents highly adaptive agriculture on steep terrain, optimized by crop rotation and rotation techniques. The crop rotation method is based on dividing the fields into first to third seasons corresponding to the types of crops and creating a "resting land" period without cultivation as a temporal cycle between the fields. On a larger spatial and temporal scale of cultivation, crop rotation is implemented to create a cycle of forest land regeneration so that the "land can self-restore" natural nutrients. The "retaining" method, also known as "abandoning the fields" (hareê talơi), is a way of "temporarily retiring" to create a period of rest for the land to regenerate the forest trees. The speed of crop rotation and rotation depends on the time of cultivation on the fields at the household level. Previously, a field was planted for two to three crops, then left fallow for more than 9 years, depending on the forest's ability to recover. When the forest area is reduced and the population increases, the fallow period must be shortened to 3–5 years. The measures of intercropping, "intensive crop cultivation," and extending the cultivation period on a field have been applied to increase the density of crop rotation. Intensive crop cultivation in the Katuic group is understood as increasing the cultivation time on a field, shortening the fallow time, investing more effort in taking care of the rice plants, weeding, chasing away wild animals, birds, rats that eat crops, etc. This can be seen as a transitional step for them to gradually switch to permanent cultivation on that field because there is no forest land to reclaim.[7]

It can be said that traditional slash-and-burn farming, which relies entirely on nature, has created a set of rudimentary labor tools and simple farming techniques "poke holes to sow seeds" is a technique that optimizes sloping terrain, minimizing erosion and leaching of seeds and nutrients. However, with this technique and tools, for thousands of years, they have not created a change in productivity and output or brought high economic efficiency, but have brought balance and stability in nature and society. However, due to the development of Vietnam in the past decades, the Katuic group has had to adapt to the limitations of traditional farming areas gradually. They were forced to convert from shifting cultivation to sedentary cultivation, gradually shifting from dry fields to wet fields, and progressively shifting from exploiting natural products to raising livestock, gardening, household economy, etc. These are compulsory settlement policies of the Vietnamese government (Nguyễn Hữu Thông et al., 2005, 394–396).

---

is how the Katuic group makes the most of the soil's natural nutrients for crops and takes advantage of the different coverage levels of crops to prevent erosion. For "new fields," the main crop is rice, in addition to intercropping some secondary crops. For second-year fields (haro halai), intercropping between rice and secondary food crops, crops (corn, cassava, potatoes, peanuts, beans, pineapple, banana, sugarcane, papaya, squash, pumpkin, taro, tobacco, cotton, etc.). For the third year of the field, intercrop with secondary food crops and flowers. This technique has created many sources of income in one field. And this has helped stabilize the Katuic group's household-level self-sufficiency economy.

[7] These are the policies of the Vietnamese government such as banning exploitation, "closing the forest", banning forest burning for farming (Nguyên Ngọc 2002; Nguyễn Tri Hùng 2005, 40–42).

Thus, the traditional "natural circular economy" in the Katuic group creates balance rather than making enormous profits, meeting internal consumption needs within the village rather than creating exchange goods. In the "natural circular economy," traditional livestock farming only plays a secondary role, not producing goods. Livestock farming on a household scale has existed for a long time, with the primary livestock being buffalo, cows, pigs, chickens, and ducks. Livestock farming is semi-natural, with free-range farming around the house or in the forest, allowing animals to mate, find food, and find shelter on their own. The notion of animism especially strongly influences livestock farming, with the primary purpose of providing animals for worship. At a higher level, the Katuic people trade and exchange goods. However, these are "primitive" exchanges (barter): exchanging naturally exploited products (honey, feathers, bamboo shoots, rattan, palm leaves, pepper, etc.) and self-sufficient products (rice, corn, woven fabrics, woven baskets, etc.) in exchange for salt, oil and production tools. This is a non-commercial exchange activity, which helps the natural circular economy of the Katuic group become more vibrant and diverse in terms of activities and resources, motivating the economy that was almost closed in the past (Bennet Bronson, 1977, 39–52; Lê Anh Tuấn, 2005; Trần Kỳ Phương, 2004; Nguyễn Bảo Đàn, 2009, 154).

In existence and development, the Katuic people in Vietnam have shifted from a primitive economy with "hunting-gathering" and "flat-and-burn" to a natural circular economy. However, they are currently facing difficulties in maintaining traditional economic activities while the reception and implementation of new economic activities have not been effective. In particular, with the policies of "closing forests," "banning slash-and-burn farming," and "banning hunting wild animals" by the Vietnamese government, the above-mentioned traditional subsistence has been almost eliminated. In addition, urbanization and modernization have enormously changed the living environment and broken the village structure and traditional social organization, leading to changes in the "natural cyclical economic" cycle, which is closed and self-sufficient. They are almost left behind and marginalized due to difficulty adapting and not keeping up with modern economic development (Đinh Hồng Hải,, 2022, 7).

Faced with many options for the current economic development model, the Katuic residents need an economic structure appropriate to their level and available resources. In the direction of economic development in ethnic minority areas, to ensure a balance between the natural and social environment, economy, and culture, it is necessary to have a solution that inherits the advantages of the "natural circular economy" and traditional indigenous knowledge to build a new economic model. Therefore, we have created an "adaptive circular economy" (ACE) model at the "microeconomic scale" of households. The knowledge and experience of the natural circular economy could be a valuable asset for the current livelihood transformation process of the Katuic residents.

5.3 Impact of Industrialization, Modernization, and Urbanization in Vietnam After Doi-Moi and the Marginalization of the Katuic People

In the worldwide impact of industrialization, modernization, and urbanization nowadays, the Ellen MacArthur Foundation gives three principles of regulations for the circular economy: (1) Preserve and enhance natural capital by controlling finite stocks and balancing renewable resource flows – for example, replacing fossil fuels with renewable energy or returning nutrients to ecosystems. (2) Optimise resource yields by circulating products, components, and materials in use at the highest utility at all times in both technical and biological cycles – for example, sharing or looping products and extending product lifetimes. (3) Foster system effectiveness by revealing and designing out negative externalities, such as water, air, soil, and noise pollution; climate change; toxins; congestion; and negative health effects related to resource use (Ellen MacArthur Foundation, 2015, 23). However, the impact of industrialization, modernization, and urbanization in Vietnam after Doi Moi[8] is strongly influenced on the habitat of ethnic minorities like the Katuic group.

5.3.1 The Impact of Industrialization, Modernization, and Urbanization in Vietnam After Doi Moi

As mentioned above, the media often sees the effects of industrialization, modernization, and urbanization (IMU) in Vietnam after Doi Moi as signs of progress. However, it's essential to acknowledge that these processes also significantly negatively impact society, culture, the environment, and individuals. Industrialization and modernization have increased demand for natural resources such as minerals, fossil fuels, and water, resulting in overexploitation and depletion. Further, industrial activities, urban expansion, and improper waste disposal have significantly impacted the environment, leading to air, water, and soil pollution. "Many emerging social problems also relate to family functions, such as same-sex marriage, birth gender selection, kidnapping and trafficking of children, poor care for the elderly and children, the lack of jobs, producing and trading for profit, virtual lifestyles, social deviancy, conflict, violence, and divorce, which exist in the entire socioeconomic system" (Teramoto et al., 2017).

Meanwhile, urban expansion often involves deforestation and the destruction of agricultural land for infrastructural purposes, causing biodiversity loss and ecosystem disruptions. Urban development also contributes to a rise in solid waste, sewage, and industrial by-products, posing health risks and contamination. In Vietnam,

[8] Doi Moi (Đổi mới) is the reform policy launched in Vietnam in 1986, translated literally as "restoration," have brought profound changes to the country.

urbanization has led to disparities in access to essential services such as education, healthcare, and economic opportunities between cities and rural areas. It drives migration to urban areas and contributes to overcrowding and poverty in the towns while causing rural work to be lost in the countryside, as we can see in Hanoi and Ho Chi Minh City's suburbs.

Gentrification due to urban development in Vietnam displaces long-term, lower-income residents, exacerbating social inequality. Besides, the blurring of work and personal life in modern work culture reduces the time for leisure, family, and self-care because the rapid pace of urbanization can overload, leading to inefficiencies and inadequate responses to human needs. While urbanization can enhance economic growth, it has widened the wealth gap, posing a threat to social stability in the mountainous areas of Vietnam. Furthermore, it affects the food security of Vietnam through land degradation and agricultural workers' displacement, labor exploitation, resource depletion, and increased poverty.

After more than three decades of IMU, societies modernized, diminishing Vietnam's traditional customs; this contributed to a more homogeneous globalized culture and a loss of cultural identity of ethnic minorities. Adopting Western values is linked to the erosion of local cultures like Katuic identities and traditions, reducing cultural diversity and unique customs. The breakdown of close-knit communities and traditional social structures leads to a loss of community. The social space has completely changed in the long-standing residential area of the Katuic people today. A society that has just left "primitive" life and immediately "clashes" with modern society, from a simple starting point, the people of the mountains and forests suddenly have to face a modern society with information overload, too many choices, the inevitable consequences are "shocks" in terms of economy, culture, and "disappointments" in terms of psychology and spirit, from the imbalance in level of awareness and reality. That gap will inevitably create difficulties in accepting new things and achievements in modern science and technology, hindering joint development, directly influenced by the traditional economy, traditional social institutions, spiritual life, and social characteristics (Trần Đức Anh Sơn et al., 2011, 69).

Therefore, addressing these challenges requires a balanced approach, emphasizing sustainable development, cultural preservation, social equity, and environmental protection. Recognizing and mitigating modernization's negative implications is crucial for establishing a more just, equitable, and sustainable future. This means adopting cleaner technologies, upholding labor rights, and ensuring equitable distribution of IMU benefits across society, which we want to apply to the Katuic region to avoid their marginalization.

5.3.2 The Marginalization of the Katuic People

The marginalization of the Katuic people is the process by which these communities are pushed to the margins of society, both economically, socially, and culturally, especially during the IMU period. The underlying cause of this situation is changing the living environment, especially after the economic opening period of Vietnam called Doi Moi in 1986. Due to the country's extensive development needs, many natural resources have been overexploited, especially forest resources. The use of wooden furniture has taken away a large amount of primary forest wood. According to the Ministry of Agriculture and Rural Development, "Vietnam has about 13.5 million hectares of forest, divided into three types: special-use forests (about 2 million hectares), protection forests (4.6 million hectares), and production forests (6.7 million hectares). Of the total 13.5 million hectares, natural forests account for about 10.3 million hectares (77%); The remaining 2.9 million hectares (23%) are planted forests. Vietnam's demand for sawn timber in 2020 increased to 7 million m^3, equivalent to 15 million m^3 of round wood; demand for fiberboard will be about 1.5 million m^3, and finger-jointed board will be about 1 million m^3 in 2015" (Trung Kiên, 2013). (A Finger-Jointed Board is a type of wood produced by joining small natural wood slats to become a larger wooden board thanks to specialized glues).

The vast majority of the remaining two million hectares of special-use forests are located in national and international reserves, which the people are not allowed to exploit. Meanwhile, for centuries, land and forests have been essential resources for Katuic communities. Therefore, the IMU and large-scale economic development projects such as logging, mining, and hydroelectric dam construction have vastly reduced almost all of their traditional subsistence. Compared to people in cities and lowlands, the Katuic people have very little access to the fruits of development and economic opportunities of the IMU. Because the Katuic communities often live in mountainous areas, far from economic centers, this makes it difficult for them to access economic opportunities such as jobs, education, and modern health services. Meanwhile, traditional teaching and healing cannot be promoted in changing conditions. They are stuck between the glamour of the IMU and a state of inescapable backwardness.

This is easily seen when traditional resources exploited from forests, rivers, streams, and slash-and-burn farming are lost; the Katuic people have become powerless in the face of rapid change and are gradually pushed to the margins of society. Lacking a voice in society, the Katuic people often have no representation or a strong voice in political and economic decisions. They are usually seen as disadvantaged and excluded from essential choices at higher levels (even though they affect their communities). Along with that is the migration and community dispersion process to serve large state projects such as hydropower, industrial zones, etc. The forced migration process due to development projects leads to the dispersion of Katuic communities; they lose community cohesion and mutual support, leading to a decline in their subsistence and traditional cultural resources. Currently, the central residential area of the Katuic group is the development site of the country's Central Electric Zone (CEZ). Once again, they are affected by land loss and changes in their living environment. The resettlement gardens are 200–500 m^2 in size, similar to the delta, leaving them misinformed about how to cultivate to earn an income to support their families. The situation of the people here is shown in our 2022 survey in Nam Đông District, Quảng Nam Province, A Lưới District, Thừa Thiên – Huế Province, and Hướng Hóa District, Quảng Trị Province, as follows (Đinh Hồng Hải,, 2022, 5).

Thus, in addition to economic marginalization, IMU has also caused the traditional cultural values, languages, and beliefs of the Katuic people to be rapidly eroded. Young people in the community tend to absorb the popular culture of the majority group and gradually lose their ethnic identity. This has caused them to assimilate with the leading group gradually. The lack of mother tongue education and cultural preservation programs further accelerates cultural assimilation, leading to the rapid disappearance of Katuic cultural characteristics. This leads to the risk of threatening the existence of the indigenous language (Adetunji, 2023), including the Katuic language group. Many indigenous Katuic languages, such as Katu, Bru, Taoih, Pako, and Kado, are at risk of being threatened because fewer and fewer people speak them, especially the younger generation. The lack of formal education and media in these languages further aggravates the situation. In this context, there is little or no institutional support for preserving and promoting Katuic languages. Without active efforts to record and revive these languages, they will face the risk of disappearing.

It is easy to see that the cultural marginalization of the Katuic group is a consequence of economic and social displacement and changes in the living environment. To address this issue, the Vietnamese government needs comprehensive policies focusing on sustainable development, cultural preservation, and empowerment of minority communities to balance environmental, social, and socioeconomic aspects and challenges of the Katuic subsistence transformation. In addition, active efforts are needed to protect and revive Katuic languages to ensure these communities are not marginalized in a rapidly changing world. It can be said that the "marginalization" of the Katuic group is not only a social problem but also a threat to the survival of Vietnamese culture and national identity based on ethnic culture. Therefore, these communities' conservation and sustainable development require intervention from many sides, including the government, NGOs, and the international community. However, it is necessary first to balance the environmental, social, economic, and cultural aspects and address the current livelihood transformation challenges of the Katuic group.

5.4 Balancing Environmental, Social, And Socioeconomic Aspects and Challenges of the Katuic Subsistence Transformation

Based on Table 5.1, we can see that the impact of IMU in Vietnam influences many aspects of living conditions, including four main elements: environment, society, economy, and culture. In theory, balancing these aspects requires a comprehensive and multi-dimensional approach, with the participation of the community, government, and relevant organizations. As stated in many official documents, such as Decision No. 432/QD-TTg on approving the Viet Nam Sustainable Development Strategy for 2011–2020 signed by Prime Minister Nguyen Tan Dung on April 12,

2012. However, starting this strategy with the people (bottom-up) is necessary to address the challenges in specific livelihood transformation because food security and traditional subsistence are the most urgent issues.

In reality, environmental change, forest resource degradation, and the construction of development projects such as hydropower dams have entirely changed people's traditional living environment and the subsistence of Katuic people. Furthermore, the intrusion of modern life into the living areas of the Katuic group has been increasing the pressure on local people. For example, local products, their primary food source, become "specialties" purchased by traders to sell in urban areas. On the other hand, people have to buy food from sources in the city while unfamiliar with buying and selling through this market, so they often suffer disadvantages when buying and selling their products. This forces them to change their traditional farming methods and get used to the "market economy," but it is beyond most people's capacity and education level.

5.4.1 Balancing the Environmental, Social, Economic, and Cultural Aspects of Katuic People

Balancing environmental, social, economic, and cultural aspects at the macro level to transform subsistence requires a comprehensive approach. This is constantly mentioned by Vietnamese media agencies and those in other parts of the world as governments worldwide. However, natural resources must be exploited sustainably, avoiding long-term damage to the ecosystem, managing forests, land, and water resources reasonably, and using environmental conservation measures to minimize negative impacts. Many developed countries apply clean technology and renewable energy, and reducing pollution is necessary to protect the environment and build strict and adequate environmental protection policies. Meanwhile, implementing such a regular and top-down approach in the whole of Vietnam is impossible, so it is necessary to have separate development policies, especially for vulnerable groups like Katuic people.

As UNDP representative in Vietnam Khalidicho Ramla emphasized, "benefits reserved for the poor often bring minimal benefit to them in the end" because the problem of falling back into poverty is always present. Of the total amount spent on the poor, "the proportion of expenditures for administration and career accounts for more than 63% of the money mobilized for poverty reduction. The expenditure for development investment accounts for only more than 36%". According to Deputy Minister Dang Huy Dong, this is because "the apparatus for managing poverty reduction programs and projects is too massive. Thus, for each poor household to access 15 million VND/year, the Fund must spend 150 million VND on the operating apparatus. That means, for every 1 VND the poor receive, 10 VND is paid to this apparatus" (Lê Nguyễn Duy Hậu, 2013).

To solve this problem, local communities like the Katuic must be involved in decision-making, especially those directly affecting their lives. However, Vietnam is autocratic, so development policies rely solely on government decisions, however top-down solutions are usually less effective. The fact that 90% of the money for the poor does not reach the poor, as mentioned above, is unacceptable. By now, people still need food and other essential needs, so small and micro-sized enterprises are currently a solution, with small business models associated with environmental protection and cultural preservation. Of course, when converting subsistence in this direction, especially from traditional agriculture to other occupations such as animal husbandry or handicrafts, Katuic people need technical support, which is related to the cultural aspect of the people.

As mentioned at the beginning of this research article, the agricultural tradition of the Katuic people in the past mainly shifted and cultivated, so they have minimal experience with modern farming and animal husbandry. Their cultural values are mostly related to the shifting and cultivation lifestyle. Therefore, preserving and promoting traditional skills is necessary to move towards sustainable development in the changing context of modern life. This transformation process requires people to acquire new techniques, technologies, and knowledge gradually. This requires educational programs that combine modern knowledge with traditional skills. Only then can programs and projects help people develop new knowledge and promote their cultural heritage, specifically their traditional experiences.

The above solutions aim to help Katuic people move towards subsistence in a non-traditional environment and adapt to social and environmental change. In other words, to balance the environmental, social, economic, and cultural aspects of the Katuic group, bottom-up subsistence transformation is the fundamental solution because it can help people participate most in this process. However, the challenge of transforming the subsistence of the Katuic people into progressive eco-innovations is a problem that has yet to be fundamentally resolved.

5.4.2 Challenge of Transforming the Subsistence of the Katuic People

The traditional lifestyle of the Katuic group is intricately connected to their natural environment, which functions as their customary habitat. Their sustenance predominantly depends on exploiting forests, rivers, and swidden agricultural methods. However, because natural resources are almost gone, finding a new livelihood sufficient to feed the population of 221,127 people, including the Bru-Van Kieu: 94,598 people, Katu 74,173 people, and Taoih 52,356 people (General Statistics of Vietnam, 2019) will be very difficult. This balance was broken when IMU invaded the residential area, completely changing their habitat.

As this habitat changes, indigenous knowledge and basic skills for subsistence, such as hunting and upland rice cultivation, are no longer viable. People are forced to transition to new occupations, such as handicrafts, services, or industry—jobs for which they lack the skills. The lack of training programs and technical assistance makes this transition even more difficult. Meanwhile, economic development at the national level, especially resource extraction and large-scale infrastructure construction, has destroyed land and natural resources (the primary sources of subsistence for the Katuic people in the past). This has severely reduced the ability to maintain traditional livelihoods with farming, hunting, and gathering methods, which are the core of Katuic culture.

According to the latest statistics of the Vietnamese government in 2019, the poverty and employment situation of the ethnic groups in the Katuic group is as follows: the Katu people have a rate of trained workers with degrees and certificates: 13.1%, the rate of workers working in the non-agricultural sector: 25.6%, the rate of poor households: 38.1%; The rate of near-poor households: 7.1%; The Taoih people have a rate of trained workers with degrees and certificates: 13.5%, the rate of workers working in the non-agricultural sector: 34.3%, the rate of poor households: 35.4%, the rate of near-poor households: 14.9%; The Bru-Van Kieu people have a rate of trained workers with degrees and certificates: 4.5%, a rate of workers working in the non-agricultural sector: 8.4%, a rate of poor households: 56%, a rate of near-poor households: 12.4% (General Statistics of Vietnam, 2019). To solve this situation, people cannot only expect funding from outside sources, but what is more necessary is a suitable livelihood conversion strategy.

5.4.3 Strategy of Subsistence Transformation

For all the reasons mentioned above, in this research, we use a bottom-up approach for subsistence transformation strategies for the Katuic group. Accordingly, the subsistence transformation strategy will start with direct workers, which in this case is the household economic model, and the effectiveness of this model will be applied to replicate on a larger scale. In the short term, the development of sustainable agricultural models, combined with livestock farming and the development of traditional handicrafts, can help improve the economic life of the Katuic people without losing their unique culture. According to my 2022 survey, people are confident that they will be able to develop their existing skills to develop their household economy. Because those skills are mostly linked to their subsistences and traditional culture (Đinh Hồng Hải, 2022, 12). In addition to the programs of the state and international organizations providing vocational training and education programs that are being implemented, the most necessary thing for the majority of Katuic people today is to have a stable income and ensure food security. Only then can they participate in new occupations and train new skills towards self-reliance and sustainable economic development.

Based on the traditional flat-and-burn livelihood that the Katuic people have had a lot of experience with from indigenous knowledge, we are building a model for converting the Katuic group's livelihood with alternative solutions from old to new as follows:

With this solution, people can take advantage of small plots of household land to develop their livelihoods and ensure food security through a balancing solution involving an adaptive circular economy and household micro-economic cells.

5.5 Balancing Solution: Adaptive Circular Economy and Household Micro-Economic Cells

Although the supporting role of the state and non-governmental organizations (NGOs) is vital, the people in the Katuic group have often been "forgotten" in the reports of international monitoring agencies such as UNDP or WB in recent years. This may be due to the small population size of the Katuic group (221,127 people), but it may also be due to the difficulty of access, as it is far from the capital and major cities. While waiting for the state and NGOs to provide technical and financial support to develop community projects, the participation of local people in small-scale economic models can create the premise to open up new opportunities.

As mentioned in Sect. 5.1, slash-and-burn agriculture is a general term for agricultural systems in which fallow vegetation is cut by hand, dried, and cleared from the field by burning before planting, while shifting cultivation is a typical farming style among the Katuic people. Based on the natural element in the natural circular economy in the traditional "slash-and-burn" of the Katuic people, we propose the "Adaptive Circular Economy" (ACE) solution. This is an attempt to materialize the ACE model based on the "natural circular economy" in the traditional "flat-and-burn" agriculture practice by the Katuic people, as mentioned in Sect. 5.1. Practicing traditional skills could help the Katuic people balance the environmental, social, economic, and cultural aspects in the process of subsistence transition with the highest efficiency.

5.5.1 Adaptive Circular Economy (ACE) solution

With the three essential contents stated in the introduction, which are "conservation and development of natural capital, optimization of resource returns, and improvement of the overall efficiency of the whole system," we have built a pilot ACE model to transform the livelihoods of the Katuic group in some of the most challenging localities. In this way, protecting land and resource rights and promoting community participation in the decision-making process are optimized. We chose the smallest unit in the ACE model, the "household," to implement. At this scale, balancing environmental, social, and economic aspects while transforming the livelihoods of the Katuic people will make it easy to adjust accordingly. More importantly, at this micro-scale, it will not require a large amount of capital, so it will not depend on programs and projects of the state as well as international organizations.

The ACE model promotes resource reuse, recycling, and regeneration within the household production area (Table 5.6). For the Katuic community, this is consistent with agricultural activities such as organic farming, using waste products as fertilizer, or making biogas. Finally, the adaptive circular economy model can help people produce goods (meat, eggs, vegetables, etc.) on a small scale in their gardens to meet household needs before being able to trade. Such activities can help people gradually access the exchange of goods within the community and exchange with the outside, creating a market for goods exchange and new economic opportunities. With such a supply and consumption chain, the goods produced will be mostly consumed locally. This is the factor that creates the sustainability of this model (Table 5.7).

Table 5.6 The Katuic habitat and subsistence

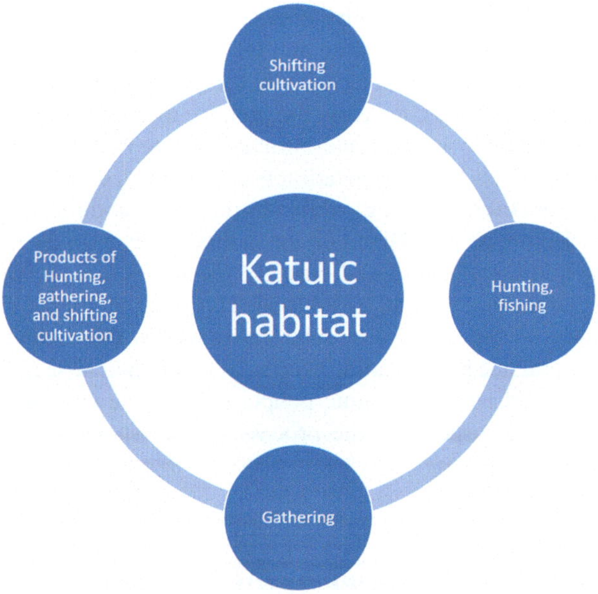

Table 5.7 From traditional to modern options for subsistence in Katuic Households

Supply	Old form	New form
Scale	Community	Household
Protein	Wildlife Hunting	Livestock raising
Starch	Slash-and-burn	Gardening, small fields
Fertility	Ashes (from flat-and-burn)	Making organic fertilizer (from livestock waste)
Vegetables	Foraging	Growing vegetables, grass, raising livestock (pigs, chickens, fish, etc.)
Consuming	Non-commercial	Post-harvest processing, output products

The problem with the ACE model is determining suitable resources, qualifications, scale, and management capacity for the family to be feasible and meet their living needs. When the process breaks the living environment in the traditional village of the Katuic region of urbanization and modernization, the strong community solidarity is also broken, and the operation of the conventional "natural circular economy" type and activities is still preserved and maintained in households - micro-economic units. Moreover, based on the traditional division of labor, which has promoted the capacity and responsibility of each family member, it is still promoting good efficiency today. Therefore, the operation of the "adaptive circular economy" model still has enough basis to rely on the household economic unit, with its inherent advantages in indigenous knowledge, experience, and diligence, as well as social relationships, which are an essential source of social capital that needs to be exploited.

5.5.2 Household Micro-Economic Cells (HMEC)

Household Micro-Economic Cells - HMEC used in this study is a microeconomic approach, especially in the context of developing countries like Vietnam. This is a small economic unit, usually a household or a small group of individuals with family ties or close relationships, engaged in production and consumption activities to sustain life and generate income. HMEC usually includes family members, sometimes extended to other members with close relationships. Economic decisions, such as production, consumption, and savings, are made within the family.

Within each HMEC, members often have different roles based on gender, age, and skills. For example, in a Katuic family, the father may be responsible for farming. At the same time, the mother takes care of the family and participates in ancillary production activities such as livestock raising or small-scale farming. Further, HMECs engage in production activities such as agriculture, handicrafts, small trade, etc. The products produced can be consumed internally or sold to generate income. Although HMECs' income is usually insignificant, it can save a lot of household expenses with "super small" investments in more profitable activities.

HMECs are the main form of the current economy of Katuic residents because they play an important role in providing all the essentials (such as labor, products, and services) on the spot. They are also the primary consumers in this "super small" economy, unaffected by the market, such as macroeconomic models. Moreover, HMECs are highly flexible in adapting to changes in economic conditions, such as when market prices change or changes in the policy environment. However, HMECs often have difficulty accessing resources such as capital, technology, and markets and face high risks due to weather, disease, or price fluctuations due to a lack of technical expertise and risk management and insurance tools. For the Katuic people, HMECs play a fundamental role in the household microeconomy before they can transition to other economic models.

Through our survey in 2022, we found that people need support to develop livestock farming (animals that have long been raised, such as pigs, chickens, buffaloes, and cows) in addition to support for production land and the renovation of rice fields. Some households want to participate in small-scale livestock farming in their households, so raising small animals such as chickens, rabbits, and porcupines is suitable. From here, we decided to select the households of Hồ Văn Khứa, A Viết Huy, Hồ Văn Dịa (A Xáp village), and Hồ Văn Khiếp (A Đền village) to pilot. These are household heads with knowledge, labor, and relatively good economic conditions. The family of Mr. A Viết Huy, the former village chief, has knowledge of raising goats, cows, chickens, etc. The family of Mr. Hồ Văn Dịa, former village Party Secretary, used to serve in the army. The family has a carpentry profession, raising fish, chickens, goats, cows, etc. The family of Mr. Hồ Văn Khứa and Viên Thị Lý graduated from Hue University of Education but have no jobs, so they stay at home to raise pigs, chickens, buffaloes, etc. The family of Mr. Hồ Văn Khứa graduated from Huế University of Agriculture and Forestry, but also has no jobs, the family has a profession of rice milling, small-scale livestock farming, etc. We hope this pilot project will help people change their thinking and awareness of production labor and strive to prosper through the labor available in the household.

Our goal is that after the breeds develop, we will multiply and share the breeds and impart knowledge in livestock farming to some other households. It is these households that, in the future, will find it easier to focus on supporting some poor households with little knowledge about animal husbandry as well as the perception of sustainable household economic development. Currently, our greatest difficulty is that there is no set of technical regulations on varieties (plants, animals) and cultivation and animal husbandry techniques suitable for the local area, climate, and soil. Hopefully, through the above pilot program, we will draw some experiences and lessons to be able to replicate them in many other households (Table 5.8).

It can be said that the subsistence transformation of the Katuic people is a challenging but also essential process in the context of rapid changes in the environment, society, economy, and culture in Vietnam today. To ensure success, cooperation between parties and the active participation of the Katuic community is crucial to ensuring that they can adapt and develop in the new environment while still maintaining the unique values of their cultural identity and traditional livelihoods. This is the reason why we chose the solution of "Adaptive Circular Economy" and "Household Micro-Economic Cells" in this research.

Table 5.8 Model of Katuic subsistence transformation with HMEC

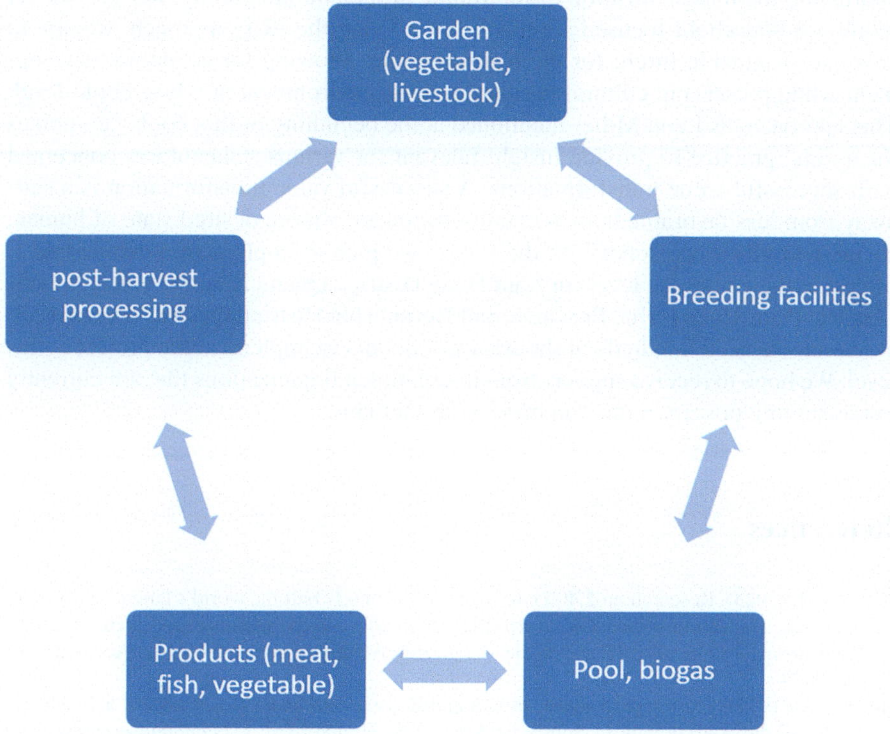

5.6 Conclusion

Based on the strategy of the Ellen Macarthur Foundation mentioned at the start of this research article, we suggested that IMU in Vietnam has the same three major problems. The food system is wasteful. It causes environmental externalities and does not produce healthy outcomes for the city population. Therefore, if we want to "advance towards a food system that is regenerative, resilient, non-wasteful, and healthier, this system would build on the technological progress of industrialization and the traditional farm system." So, "setting the true price of resources and natural capital, pricing externalities, shifting taxes away from labor, and removing environmentally harmful subsidies could make this vision a reality" (Ellen Macarthur Foundation, 2015, 70). Therefore, to successfully implement the circular economy and improve the subsistence of the Katuic people, it is essential to provide skills training in sustainable agriculture, waste management, and small and micro-enterprise development.

While the government and international organizations have supported infrastructure improvements such as roads, communication networks, and access to clean water, the participation of local communities in the circular economy is still lacking in this area. Balancing environmental, social, economic, and cultural aspects while addressing the marginalization of the Katuic group and integrating the ACE model requires a household-focused approach. By piloting the ACE approach, we aim to create a sustainable future for the Katuic people, allowing for economic development while preserving cultural heritage and the environment at a household level. This approach, as Leon Miller mentioned at the beginning of this book, "combines theory and practice to provide insight relevant for various stakeholders concerned with successful value transformations. A successful value transformation is a shift away from less favorable human conditions toward a more desired state of human, social, and global existence". In the future, we plan to implement a detailed ACE project at the commune level (in Nam Đông District, Quảng Nam Province, A Lưới District, Thừa Thiên – Huế Province, and Hướng Hóa District, Quảng Trị Province). However, we need hundreds of thousands of dollars to implement the project at this level. We hope to receive support from international organizations that are currently implementing poverty reduction projects in Vietnam.

References

Adetunji, J. (2023). Research on 2,400 languages shows nearly half the world's language diversity is at risk. *The Conversation UK*, April 2023. https://theconversation.com/research-on-2-400-languages-shows-nearly-half-the-worlds-language-diversity-is-at-risk-204014. Accessed 18 Oct 2024.

Bennet, B. (1977). Exchange at the upstream and downstream ends: Notes toward a functional model of the coastal state in Southeast Asia. In K. Hutterer (Ed.), *Economic exchange and social interaction in Southeast Asia: Perspectives from prehistory, history and ethnography* (Vol. 1977, pp. 39–52). Center for South and Southeast Asia Studies, University of Michigan.

Đinh Hồng Hải. (2022). 'Forever marginalized': Engaged research on Katuic's ethnic minorities in Vietnam. *Engaged Research Grant*. Wenner—Gren Foundation, Grand Number: Gr. ERG-7.

Đỗ Đình Sâm. (1994). *Shifting cultivation in Vietnam: Its social, economic, environmental value relative to alternative land use*. International Institute for Environment and Development (IIED). Forest Science Institute of Vietnam (FSIV).

Ellen MacArthur Foundation and McKinsey Center for Business and Environment. (2015). Growth within: A circular economy vision for a competitive Europe. https://www.ellenmacarthurfoundation.org/growth-within-a-circular-economy-vision-for-a-competitive-europe. Accessed 29 Aug 2024.

Erni, C. (Ed.). (2015). Shifting cultivation, livelihood and food security: New and old challenges for indigenous peoples in Asia. In *Shifting cultivation, livelihood and food security new and old challenges for Indigenous peoples in Asia*. FAO, IWGIA and AIPP.

European Parliament. (2023). *Circular economy: Definition, importance and benefits*. https://www.europarl.europa.eu/topics/en/article/20151201STO05603/circular-economy-definition-importance-and-benefits#:~:text=What%20is%20the%20circular%20economy,cycle%20of%20products%20is%20extended. Accessed 28 Aug 2024.

General Statistics of Vietnam. (2019) (Tổng cục Thống kê, www.gso.gov.vn). See more detail in: https://www.gso.gov.vn/du-lieu-dac-ta/2019/12/htcttkqg-ty-le-lao-dong-da-qua-dao-tao/. Accessed 21 July 2024.

Huy Hoàng. (2017). Panoramic view of Vietnam's natural forest since 1945. *Journal of Environment*. https://tapchimoitruong.vn/thien-nhien-va-moi-truong-viet-nam. Accessed 19 Oct 2024

Kleinam, P., Pimentel, D., & Bryant, R. B. (1995). The ecological sustainability of slash-and-burn agriculture. *Agriculture, Ecosystems and Environment, 52*(2), 235–249.

Lê Anh Tuấn. (2004). Traditional Katu economy and some issues raised (case study of Avuong commune, Tay Giang, Quang Nam). In *Scientific information* (pp. 91–105). Central Institute of Cultural and Artistic Research in Hue City. March issue.

Lê Anh Tuấn. (2005). Ethnic minorities and cross-border trade activities in the central region (survey results in Quang Tri, Thua Thien and Quang Nam). In *Scientific information* (pp. 171–193). Vicas Hue.

Lê Nguyễn Duy Hậu. (2013). Where did the money for poverty reduction go? In *Pháp luật Việt Nam*. https://baophapluat.vn/tien-xoa-doi-giam-ngheo-da-di-dau-post174619.html. Accessed 28 Aug 2024

Nguyên Ngọc. (2002). Some issues on land, forests and villages in the mountainous areas of Quang Nam. In *Ngok Linh (special topic of research and writing on mountainous areas and the Central Highlands)* (Vol. 03, pp. 5–12). Da Nang Publishing House—Center for Social Sciences and Humanities, Da Nang University.

Nguyễn Hữu Thông Ed, Lê Anh Tuấn, Trần Đình Hằng, Nguyễn Phước Bảo Đàn, Tôn Nữ Khánh Trang, Trần Đức Sáng, & Trần Thanh Hoàng. (2005). *The village culture of mountainous areas in Central Vietnam: –Its traditional values and historical changes (Data drawn from the Mountainous areas of Quangnam province).* Thuận Hóa Publisher.

Nguyễn Phước Bảo Đàn. (2009). From the "salt road" identifying the historical back-and-forth exchange network in Central Vietnam. In *Perception of Central Vietnam—10-year journey of approach* (pp. 151–217). Thuan Hoa Published.

Nguyễn Tri Hùng. (2005). Cultural economy of mountainous villages in Quang Nam. *Culture and Arts Magazine, 03*, 40–42.

Peters, W. J., & Neuenschwander, L. F. (1988). *Slash and burn. Farming in the third world forest.* University of Idaho Press.

Stefan, H., & Lindsey, N. (2013). Slash-and-burn agriculture, effects of. In S. A. Levin (Ed.), *Encyclopedia of biodiversity* (Vol. 6, 2nd ed., pp. 551–562). Academic.

Teramoto, M., Nguyen Duc Chien, Misaki Iwai, & Bui The Cuong. (2017). *The Vietnamese family during the period of promoting industrialization, modernization and international integration.* IDE-JETRO.

Trần Đức Anh Sơn, Nguyễn Hữu Thông, Lê Anh Tuấn, Nguyễn Phước Bảo Đàn, Trần Đức Sáng, Bùi Thị Thuận, & Nguyễn Đăng Châu. (2011). *Research on the impact of urban development on the cultural and social life of the Katu community in Hoa Vang district, Da Nang city.* Project report of Da Nang Institute for Socio-Economic Development Research.

Trần Kỳ Phương. (2004). Initial study of the geo-history of the Champa kingdom in Central Vietnam: With special reference to the "riverside exchange system" of the Thu Bon river basin in Quang Nam. In *Scientific information* (Vol. 3, pp. 48–49). Vicas Hue.

Trung Kiên. (2013). Natural forests under pressure from demand for wood. *ThienNhien.net*. https://daln.gov.vn/rung-tu-nhien-chiu-suc-ep-tu-nhu-cau-su-dung-go.23238.news. Accessed 9 July 2024.

Vosti, S. A., & Witcover, J. (1996). Slash-and-burn agriculture- household perspectives. *Agriculture, Ecosystems and Environment, 58*(1), 23–38.

Chapter 6
Eco-Innovations and Circular Agriculture Development in Vietnam

Ho Ngoc Son

Abstract Vietnam's agricultural sector is crucial to its economy and food security, but it faces sustainability challenges, including resource depletion, pollution, and climate impacts. Circular agriculture, which minimizes waste and maximizes resource efficiency, offers a path to sustainable agricultural transformation. Eco-innovations which include technologies and practices that support circular principles by reusing resources, reducing waste, and regenerating ecosystems are key to achieving this. This chapter explores eco-innovations in Vietnam's agricultural landscape, including rice-fish farming, biogas production, composting, precision farming, and aquaponics. The benefits of these innovations span environmental, economic, and social dimensions, but challenges such as limited access to technology and knowledge gaps hinder their widespread adoption. The paper concludes with recommendations to support Vietnam's transition to circular agriculture through eco-innovations.

Keywords Circular agriculture · Eco-innovations · Sustainable farming · Agricultural waste recycling · Resource efficiency · Biogas production · Rice-fish farming · Climate-resilient agriculture

6.1 Introduction

Agriculture is critical to Viet Nam's economic development, food security, and livelihoods. It now contributes about 14 percent of gross domestic product (GDP) and about 38 percent of employment. The sector played a major role in reducing poverty to less than 6 percent, bringing about almost zero food insecurity, maintaining social stability, and with many agricultural commodities, including rice, coffee, cashew, pepper, fruits, vegetables, and seafood entering global markets, earning the country

H. N. Son (✉)
Thai Nguyen University of Agriculture and Forestry, Thái Nguyên, Vietnam

S. Patnaik et al. (eds.), *Circular Economy and Sustainable Value Creation through Eco-Innovation*, Circular Economy and Sustainability,
https://doi.org/10.1007/978-3-031-88516-7_6

over US$48 billion per year in export revenues in 2021 (World Bank, 2022). However, traditional agricultural practices have led to resource depletion, soil degradation, and water pollution, particularly in regions like the Mekong and Red River Deltas where intensive rice cultivation dominates. Rice production accounts for almost half of GHG emissions in the agricultural sector in Viet Nam and 75% of total national methane emissions (MONRE, 2020). These high levels of emissions can be explained by inefficient irrigation, fertiliser application, energy use, and poor management of rice residue and livestock by-products. Given the scale and importance of the agriculture sector in Viet Nam, including its environmental impacts, its contribution to GHG emissions, and its vulnerability to climate change, it makes infinite sense to leverage the potential of circular models at scale.

In addition, the total amount of residue and agricultural by-products from staple and industrial crops in Viet Nam is estimated at 95 to 98 million tons per year, including 52 million tons of straw and husk from rice production. Viet Nam's livestock population generates close to 72 million tons of solid waste and over 76 million tons of liquid waste per year. When left untreated, this waste can have significant negative impacts on water and soil, in turn affecting crop quality. The burning of waste and agricultural residues in particular during harvest season, contributes to worsening local air quality with disastrous impacts on health, especially for the most vulnerable, the children and elderly. As Vietnam's population grows and climate change impacts intensify, there is an urgent need to shift from linear to circular agricultural models that reuse resources, reduce waste, and protect ecosystems. Circular agriculture is a way to farm sustainably by adopting scientific advances, innovations, and new technologies. The vital components of circular agriculture are integrated crop, livestock, fishery, agroforestry, and waste and water recycling (Ellen MacArthur Foundation, 2016).

Several circular agriculture practices exist in Vietnam, for examples, VAC or Garden -Pond -Barn and VACR or Garden -Pond -Barn- Forest systems to make use of biomass; models of collecting agricultural waste such as tree trunks, straw, and rice husks; a bioaquatic model; and an organic farming model. In cultivation, the by-products of peanut shells, corn stalks, rice straw, cassava stalks, soybean hulls, firewood from perennial plants, and firewood from annual plants are used. Straw is also used as food for cattle, biological padding, mushroom cultivation, and for composting as traditional organic fertilizer and burning. Regarding livestock, by-products are used in traditional composting, biogas production, microbial manufacturing biological padding and for commercial purposes such as earthworm farming. In aquaculture, the collection rate is more than 90%; by-products are used for the extraction process of biological compounds; food products, such as fish oil; and producing organic fertilizer (Manh & Pham, 2020).

Eco-innovations are at the heart of circular agriculture's evolution, enabling farmers to adopt sustainable methods and technologies (Praveen et al., 2024). These include innovative ways to recycle agricultural waste, integrate aquaculture with crop production, and use data-driven solutions for water and pest management. By adopting circular principles and eco-innovations, Vietnam has the potential to sustainably boost productivity, reduce agricultural waste, and ensure food security.

This chapter discusses how circular economy practices, combined with eco-innovations, can drive sustainable agriculture in Vietnam, analyzing specific innovations, the benefits they provide, and the challenges in their adoption.

6.2 Evolution of Circular Agriculture Practices in Vietnam

Circular agriculture in Vietnam has spanned centuries, rooted in its agrarian society, and shaped by the country's distinct geography, culture, and resource constraints. From ancient techniques that promoted harmonious relationships between nature and farming to modern technological advancements, circular agriculture in Vietnam has evolved in profound ways (Son & Tuan, 2022).

6.2.1 Traditional Circular Agriculture

Circular agriculture in Vietnam has its origins in ancient practices that were deeply tied to the land and local ecosystems. Farmers worked in close harmony with nature, often out of necessity, as resources were limited and rural communities were isolated from industrial advancements. The earliest forms of circular agriculture emerged in the river deltas, where communities learned to manage and reuse resources to support their farms. The Mekong and Red River Deltas became hubs for unique farming systems that optimized water, land, and nutrient cycling. Techniques such as crop-livestock integration and nutrient recycling were fundamental. The rice-fish farming system, known as Lúa-Cá, was one such method where rice paddies doubled as fish habitats. Fish provided natural pest control and fertilization, creating a symbiotic relationship that enhanced soil fertility and productivity without reliance on external inputs. Additionally, the VAC system (Vườn-Ao-Chuồng) emerged, integrating gardens (Vườn), fish ponds (Ao), and animal barn (Chuồng) into a self-sustaining unit. Garden waste would feed livestock, livestock manure would fertilize gardens and ponds, and pond plants would become feed. This closed-loop system minimized waste, optimized resources, and offered farmers diverse sources of income. These traditional systems were built on localized knowledge passed down through generations, proving to be resilient and adaptable to changing conditions (Son et al., 2021).

6.2.2 Colonial Era and Agricultural Shifts

During the colonial era in the late 19th and early 20th centuries, Vietnam saw the introduction of European agricultural techniques. French colonists promoted cash crops like rubber, coffee, and tea for export, which shifted some agricultural focus from subsistence to commercial farming. Monoculture practices were introduced, and with them came the first signs of dependency on external inputs like fertilizers and pesticides. These shifts disrupted traditional farming practices. Farmers in areas transformed for cash crops had fewer opportunities to practice circular agriculture, as monoculture did not support integrated systems like rice-fish farming or the VAC model. However, in regions where traditional practices persisted, circular agriculture maintained its role, particularly in rural villages less influenced by colonial policies.

6.2.3 Reconnecting with Traditional Agriculture

Following Vietnam's reunification in 1975, the country's agricultural sector faced enormous challenges. Food shortages and low productivity plagued rural communities, and the government prioritized efforts to increase agricultural output to achieve food security. Initially, this meant adopting Green Revolution techniques, such as high-yield crop varieties and synthetic fertilizers. While these measures helped increase food production, they also created environmental and economic challenges, including soil degradation, water pollution, and the rising cost of agricultural inputs. Recognizing the importance of sustainable and resource-efficient farming, the government began to encourage traditional practices again in the 1980s and 1990s. Policies promoted integrated systems, and the VAC model saw a resurgence. The Vietnamese government established local training centers, and farmers received education on traditional and integrated farming practices, highlighting the benefits of nutrient cycling, waste reduction, and resource optimization (Son et al., 2021).

6.2.4 The Rise of Environmental Awareness and Global Influence

Entering the 21st century, Vietnam faced mounting environmental concerns, such as soil erosion, water pollution, and the impact of climate change. Coastal areas in particular were vulnerable, as rising sea levels threatened the fertile deltas that were the backbone of Vietnamese agriculture. In response, the government, local organizations, and international partners increasingly supported circular agriculture as a way to build climate resilience. Vietnam began collaborating with international organizations such as the FAO and UNDP, which helped introduce sustainable agricultural practices and modern technologies to enhance traditional circular systems. At the same time, Vietnamese scientists and agricultural institutions developed new crop varieties and techniques tailored to local climates and ecosystems. Research into high-yield, pest-resistant rice strains allowed farmers to reduce pesticide use, which complemented traditional systems like Lúa-Cá, further reducing environmental impact.

The early 2000s also saw the integration of biogas technology on farms, where small-scale digesters converted livestock manure into biogas, an alternative energy source. This technology enabled households to power cooking and lighting while producing nutrient-rich slurry to fertilize fields, further closing resource loops. The reintroduction of these circular practices was well received by farmers, who found that reducing dependency on external inputs lowered costs and increased productivity (Son et al., 2021).

6.2.5 Technological Integration and Innovation

Today, Vietnam stands at the forefront of merging traditional circular agriculture with cutting-edge technology, leading to an innovative hybrid of old and new. Advances in precision agriculture, IoT-based irrigation, and data analytics are helping farmers optimize resource use, conserve water, and reduce environmental impacts. For example, drones and remote sensing are used to monitor crop health, soil moisture, and pest levels. This data-driven approach has minimized the need for fertilizers and pesticides in integrated systems, making them more environmentally friendly and cost-effective. In aquaculture, automated aerators and feeders powered by solar energy maintain water quality and fish health, increasing productivity in rice-fish farming.

Digital platforms and blockchain technology are revolutionizing supply chain transparency, allowing consumers to trace the origin of their food. This transparency is especially valuable for Vietnamese farmers producing organic or sustainably farmed products, as it enables them to access premium markets and gain better prices. E-commerce platforms also help rural farmers reach urban consumers, increasing the demand for produce grown with circular principles.

Biotech advancements are further supporting circular agriculture. Improved rice varieties that are pest-resistant and require less water are now being cultivated, allowing traditional rice-fish systems to thrive under harsher climate conditions. Probiotics for livestock, an alternative to antibiotics, enhance nutrient absorption and reduce waste, making livestock farming cleaner and more efficient.

6.2.6 The Future of Circular Agriculture in Vietnam

As Vietnam continues to grow and industrialize, the evolution of circular agriculture practices faces both challenges and opportunities. Climate change, urbanization, and global economic pressures will shape the path forward, pushing Vietnam to find innovative solutions to sustain circular practices while meeting modern agricultural demands. Vietnam's rich history of circular agriculture, rooted in deep cultural values, is not just a relic of the past but a blueprint for the future. By adapting traditional systems with technology, Vietnam is pioneering resilient, sustainable agriculture models that could serve as examples for other countries. The Vietnamese government's continued support for sustainable agriculture, combined with the ingenuity of local farmers and the aid of global partners, will be crucial in this evolution.

6.3 Traditional Circular Agriculture Practices

Circular agriculture models have been practiced by local communities in Vietnam for generations. These communities, which also include ethnic minority groups like the Hmong, Dao, Tay, and Thai, live in mountainous and remote areas where they rely on resilient, low-input agricultural practices that sustain both their environment and their livelihoods. These traditional practices integrate crops, animals, and forest resources, creating closed-loop systems that maximize productivity while safeguarding the natural environment (Son et al., 2021). Many traditional circular agriculture practices were effective in sustainable waste management, renewable energy production, pollution reduction, and greenhouse gas emission reduction. These traditional circular systems are celebrated for their resilience and sustainability, providing Vietnamese communities with food security, reducing environmental impact, and building community resilience in the face of climate change.

6.3.1 Garden-Pond-Barn (VAC System)

A prominent Vietnamese system called VAC (Vườn-Ao-Chuồng) integrates gardens (Vườn), fish ponds (Ao), and barn (Chuồng). Farmers plant fruits and vegetables, raise fish in ponds, and keep pigs. Waste from pigs is used to fertilize the garden and fish pond. In return, aquatic plants from the pond serve as food for livestock, and nutrient-rich pond water fertilizes garden plants. The VAC model has been widely applied in Vietnam since the 1980s and considered the simplest form of circular agriculture, in which: (i) The garden is the cultivation activity. In the garden, farmers plant a combination of many kinds of crops and trees with different layers to take advantage of solar energy and soil nutrients. Normally, farmers grow vegetables, beans, some spices, and medicinal plants in the garden's corner. If farmers have a large garden, they can grow timber combined with fruit trees as a fence to cover the wind; (ii) In the pond, people usually raise many types of fish at different layers. By doing so, farmers can take advantages of food sources from the garden and livestock waste for fish; (iii) Barn is a place where cattle (pig, goat, cow, and buffalo) and poultry were raised (Ninh et al., 2023).

VAC has created an integrated agricultural production model, linking farming with animal husbandry, minimizing waste, and following nature. The VAC model has been improved by the development level of agricultural production as well as the ecological conditions of each territory across the country, that is "Garden - Pond - Barn – Biogas" (VACB); "Garden - Pond – Barn – Forest" (VACR) in mountainous provinces; "Garden - Pond – Lake" (VAH) in the Central provinces. Implementing the VAC model not only brings high economic efficiency but also helps to reduce GHG emissions. Especially, the VACB model has helped to manage agricultural waste, rationally use agricultural waste and by-products as fertilizer to return soil fertility, safely treat animal waste, create renewable energy, create renewable energy sources of fuel for daily life, fight against environmental pollution and contribute to reducing greenhouse gas emissions and reducing greenhouse effects. The VAC model in Vietnam was initially small, to contribute to food security, hunger eradication, and poverty alleviation. The VAC model has gained significant popularity and widespread adoption across the country. It has undergone improvements and demonstrated flexibility in its applications, benefiting numerous households, farms, cooperatives, and enterprises in organizing production and business activities.

6.3.2 Intercropping and Crop Rotation

Intercropping and crop rotation have been practiced in Vietnam for generations. By rotating crops and planting complementary species together, farmers improve soil fertility, control pests naturally, and reduce the need for chemical inputs. These traditional techniques contribute to biodiversity and soil health, two essential aspects of circular agriculture. In the highland areas, ethnic groups such as the Hmong and Dao practice intercropping, a method that involves planting multiple crops together to maximize land productivity and support soil health. Maize, beans, and vegetables are commonly grown in the same fields. Beans, being nitrogen-fixing plants, enrich the soil and support the growth of other crops without the need for synthetic fertilizers (Son et al., 2021). Crop rotation is also practiced, allowing the soil to regenerate by alternating between nutrient-intensive crops like maize and nitrogen-fixing plants, thereby reducing pest and disease pressure naturally.

6.3.3 Integrating Livestock and Crop Production

In many Vietnamese communities, livestock is an integral part of the farming ecosystem. Pigs, chickens, buffaloes, and cattle are commonly raised alongside crops, creating a symbiotic relationship between animals and agriculture. Livestock manure is collected and used to produce organic compost, which is then applied to crop fields as a natural fertilizer. This practice recycles nutrients within the system, reducing dependence on synthetic fertilizers and enhancing soil fertility. For example, the Dao people in northern Vietnam raise buffaloes in areas where rice is grown, collecting the manure for rice field fertilization and plowing (Son et al., 2019). The animals not only contribute nutrients but also help with field preparation, reducing labor for farmers. Some communities take integration even further by using crop residues to feed livestock, which in turn provide more manure for the fields. This closed-loop system ensures that few resources are wasted, with each component benefiting the others. For instance, rice bran left over after harvest is often fed to pigs and chickens, while crop stalks become feed for buffaloes. This synergy maximizes resource use, contributing to a more sustainable and resilient farming system.

6.3.4 Agroforestry Systems

Agroforestry is a traditional practice among many ethnic groups in Vietnam, particularly those living in upland areas. Agroforestry, defined as tree planting in combination with crops or pastures, is an integral part of circular agriculture. Tree planting can help restore biodiversity in agricultural landscapes, while increasing soil fertility by enhancing the accumulation of organic matter from de- caying nature. Agroforestry can also make agriculture more circular by reducing the dependency on chemical fertilizers and pesticides. As agroforestry reduces the need for inputs, it is more accessible for female farmers who often have less financial resources and limited access to credit, which can provide new opportunities for women's empowerment in the rural economy (Helgason et al., 2021). In addition, the collection of firewood and fodder is primarily the task of women in rural areas around the world. Agroforestry makes these products available on the farm and thus reduces the time spent by women on such activities, which can contribute to their empowerment. Many ethnic minorities practice agroforestry, where trees are grown alongside crops to provide shade, reduce soil erosion, and improve biodiversity. For instance, the Tay and Thai people in the Northern Vietnam plant fruit trees such as longan and jackfruit on sloped lands, which help stabilize the soil. These trees also create a microclimate that protects crops from extreme weather and provides families with additional food sources. Agroforestry systems reduce reliance on chemical fertilizers by promoting natural nutrient cycling through leaf litter decomposition and enhancing overall farm resilience.

6.3.5 Aquaculture-Rice Integration

In many areas, farmers practice integrated rice-fish farming and rice-shrimp farming. By combining aquaculture with rice cultivation, these communities create a mutually beneficial ecosystem that aligns closely with the principles of circular agriculture.

In Vietnam's Mekong and Red River Deltas, farmers often cultivate rice and fish together in the same paddy fields. Rice paddies are modified to hold fish, which feed on pests and provide natural fertilizer for the rice plants through their waste. This symbiosis reduces the need for chemical pesticides and fertilizers, boosts biodiversity, and provides farmers with a source of protein and income from both rice and fish. In the Mekong Delta, the rice-fish farming system is a longstanding example of circular agriculture. Farmers cultivate rice alongside fish, where fish feed on pests and provide natural fertilizer to the rice plants. This symbiotic relationship reduces the need for pesticides and chemical fertilizers, improving soil quality and water retention. The rice-fish system enhances biodiversity and creates an additional source of income for farmers.

The "rice-shrimp" model has been applied since the early 2000s in Mekong Delta, while the "rice-fish" model has been implemented in the lowland and flooded areas in the Red River Delta. These models have been evaluated as the most relevant ones in the coastal saline areas. In these models, shrimp or fish is raised in the rice field, and their excrement and leftover food are used as a nutritional supplement for rice. On the contrary, when the rice is harvested, shrimp are released into the field, and the stubble and scattered rice become a source of food for shrimp and fish. In these models, crops and livestock do not need pesticides and antibiotics. Therefore, they both produce quality products and protect the environment. The models of "rice-shrimp" and "rice-fish" can create income that is as higher as 5 to 10 times as rice only (Ninh et al., 2023).

The rice—shrimp and rice—fish farming system create an ecological balance and a safe environment with mutual advantages for both crops and livestock. It also restricts pests for both rice and livestock through the practice of virtuous crop rotation, thereby creating a unique system of mutual benefit. Additionally, it enhances the ability to break down and remove toxic elements due to alternating between salty and freshwater regimes and the presence of plant flora. This system reduces production costs by minimizing the need for fertilizers due to the presence of residual organic sources that enhance plant growth and by decreasing the use of pesticides, thanks to reduced pest populations through crop rotation. It also curtails environmental pollution, ultimately resulting in clean, organic products that benefit human health, provide valuable export commodities, increase incomes for farmers and businesses involved in the production-to-consumption chain (Quy, 2024).

6.3.6 Integrated Rice-Duck Farming System

The model of "rice-duck farming" has integrated animals, plants and feed together to tap the resources in rice paddies. In terms of ecological mutualism, rice and duck make a perfect match. This mode has made full use of these natural resources of the rice paddies, light, heat, water, soil and gas, not only achieving cheering rice and duck harvests at low costs but also facilitating the virtuous circle of the ecosystem of the rice paddies. With this model, ducks are raised when double or single cropping rice is planted in double or single seasons. Three-week-old ducklings are introduced in a rice paddy about 20–30 days after the seedlings have been planted. Feeds have to be released into the Felds rationally and carefully. The ducks are left in the field for 60–80 days and nights. Once the rice plants form ears of grain, they have to be taken away for sale. All the processes shall be arranged in a proper order of time. The "rice-duck farming" upgrades Vietnam's traditional agriculture to the matching degree of the habitat featuring intergeneration, mutual promotion and inter-restriction of the organisms. Realizing planting and breeding both in the field, it has broken through the traditional method of duck feeding. Fulfilling ecological mutualism, it has also opened up a new way to integrate rice cropping with animal breeding amidst the thundering changes of the agricultural management system (Quy, 2024).

6.3.7 Composting and Recycling Organic Waste

In many rural areas, animal manure is collected and applied to fields as a natural fertilizer. This practice minimizes the need for synthetic fertilizers, which are costly and often harmful to the environment. By using animal waste as a nutrient source, farmers promote soil health and reduce agricultural pollution, which aligns well with circular agricultural principles. In ethnic minority communities, organic waste is rarely discarded. Kitchen scraps, animal manure, and crop residues are composted to produce nutrient-rich organic fertilizer, which is then applied to fields. The Hmong and Dao, for instance, create compost piles near their homes where organic waste decomposes into rich humus that improves soil quality. This process not only reduces waste but also enriches the soil, helping to maintain healthy crops year after year. This recycling of organic matter is central to circular agriculture. By using compost instead of chemical fertilizers, farmers reduce their environmental footprint and promote the health of their soils. Composting is also a flexible solution that can be adapted to various types of waste, from food scraps to rice husks, allowing households to effectively manage waste and enhance farm productivity.

6.3.8 Case Studies

6.3.8.1 Case Study 1: Tay Farmers and Agroforestry in the Upland Areas

The Tay people, living primarily in the northeastern mountains, have developed complex agroforestry systems. They plant a variety of crops alongside trees, such as jackfruit and longan, which provide both economic and ecological benefits. This system prevents soil erosion on steep slopes, helps maintain soil fertility, and increases overall productivity. The trees act as windbreaks and create a favorable microclimate for crops, reducing vulnerability to extreme weather. The Tay also incorporate livestock, such as chickens and buffalo, into their agroforestry systems. Animal manure is composted and returned to the fields as fertilizer, creating a closed nutrient loop that enriches the soil without reliance on chemical inputs. This practice exemplifies a sophisticated, circular approach that utilizes the land's full potential while maintaining ecological balance.

6.3.8.2 Case Study 2: The Thai Community's Integrated Rice-Fish Farming System

The Thai community, especially in the Red River Delta and lowland regions, practices integrated rice-fish farming. In this system, fish (usually carp or tilapia) are raised within flooded rice paddies. Fish waste naturally fertilizes the rice, while the rice paddies provide a habitat for fish. The fish help control weeds and insect populations, reducing the need for chemical pesticides and fertilizers. This integration improves farm productivity and resilience, providing families with both staple rice and a secondary protein source. By raising two complementary food sources in a single system, the Thai create a self-sustaining and efficient method that enhances food security and contributes to income diversification.

6.3.8.3 Case Study 3: Dao Ethnic Group and Composting of Organic Waste

The Dao community traditionally relies on composting organic waste, including crop residues and animal manure, as a key practice for soil fertility. Near their homes, Dao households construct composting pits where organic materials decompose into nutrient-rich humus. This compost is then applied to their terraced fields, supporting a regenerative cycle that enriches the soil over time. The Dao's composting practices reflect a deep understanding of the value of organic matter in sustaining crop health. Instead of using chemical fertilizers, which can degrade soil quality, the Dao rely on this closed-loop nutrient cycle, creating a sustainable system that preserves soil health and reduces reliance on external inputs.

6.4 Eco-Innovations and Circular Agriculture Development in Vietnam

6.4.1 Improved Composting and Biofertilizers

While traditional composting methods remain prevalent, development organisations and agricultural programs have introduced advanced composting methods that speed up decomposition, such as the use of composting bins and natural enzymes. These innovations enable farmers to produce compost faster and on a larger scale, allowing for more consistent soil enrichment. This practice not only increases crop yields but also reduces the amount of waste and improves soil health (Selvan et al., 2023).

Some farms use biochar, a carbon-rich fertilizer made from organic waste, which improves soil health and sequesters carbon. Innovation aspect of modern composting techniques is the use of aerobic composters and microbial inoculants that speed up decomposition and enhance the nutrient profile of compost. Biochar technology, adapted to local crop waste, increases carbon sequestration and soil water retention. Composting reduces waste and reliance on chemical inputs, improves soil health, and contributes to carbon sequestration. Biofertilizers support sustainable soil productivity, promoting long-term agricultural resilience.

Vermicomposting, the use of earthworms to process organic waste into nutrient-rich compost, has gained traction among smallholder farmers and urban households. Worms, also known as cinnamon worms, belong to the group of fecal worms and are typically found in environments rich in decomposing organic matter. They are a preferred and popular food source for animals like ducks, chickens, geese, ornamental birds, and fish. Cinnamon worms possess the advantageous ability to break down organic matter. One ton of cinnamon worms can effectively decompose up to 80 tons of organic waste. Therefore, worm farming serves as a method for treating domestic organic waste.

Black soldier fly farming is also widely employed in waste management within animal husbandry. This approach helps control house flies and transforms organic waste into valuable products such as organic fertilizer. Black soldier flies are insects commonly found in outdoor environments near livestock or areas with decomposing organic matter, including animal waste. Black soldier flies are a rich source of nutrients suitable for animals such as pigs, chickens, ducks, and serve as an exceptional live food option for specialty animals like shrimp, crabs, fish, and frogs. In Vietnam, black soldier flies are successfully utilized as animal feed in various localities.

6.4.2 Improved Rice-Fish Farming Systems

Rice-fish farming, a traditional method in the Mekong Delta, has evolved as a practical circular agriculture model. In this system, fish are raised in flooded rice paddies, where they help control pests by feeding on insects and weeds. The fish waste serves as a natural fertilizer, enhancing soil fertility and reducing the need for chemical inputs. Innovation aspect of the modern rice-fish systems is the incorporation of improved fish varieties and irrigation management, making the method more efficient and sustainable. Integrating sensors and data analysis tools helps farmers optimize water use, reducing waste and emissions. Rice-fish farming reduces chemical pesticide and fertilizer use, lowers methane emissions, and provides farmers with an additional income stream from fish sales.

6.4.3 Biogas Production from Livestock Waste

Biogas production from livestock waste has become an important eco-innovation in Vietnam, especially in regions with high livestock populations, like the northern provinces. Livestock manure is processed in biogas digesters, which capture methane and convert it into biogas for cooking and heating, while the by-product, a nutrient-rich slurry, is used as a natural fertilizer. Innovation aspect of new biodigester models is the improvement in methane capture efficiency and slurry composition, enhancing both energy production and soil fertility. Portable, small-scale biodigesters have been introduced to accommodate smaller farms, making biogas more accessible. This innovation reduces methane emissions from manure, decreases dependency on fossil fuels, and provides farmers with a renewable energy source, cutting energy costs.

6.4.4 Precision Farming and Digital Agriculture

Precision farming technologies are emerging in Vietnam to enhance resource efficiency. Farmers use drones, satellite imagery, and soil sensors to monitor crop health and manage inputs precisely. For example, smart irrigation systems help optimize water use by tracking soil moisture, while drones enable targeted pest control. Innovation aspect of precision tools and data-driven platforms is the real-time monitoring of crops, allowing for resource-efficient farming practices. Mobile platforms that provide farmers with weather forecasts, market prices, and agricultural advice make circular practices more accessible. Precision farming reduces input waste, improves yields, and enables farmers to adapt more easily to climate variability, making agriculture more sustainable and profitable.

6.4.5 Aquaponics in Urban and Suburban Areas

In cities like Ho Chi Minh and Hanoi, small-scale aquaponics systems are being adopted to grow vegetables and fish in a closed-loop system. The nutrient-rich water from fish tanks is recycled to fertilize plants, which filter the water before it returns to the fish tanks. This approach reduces water and nutrient waste, creating a circular system suited for urban agriculture. Innovation aspect of compact aquaponics systems are designed for small spaces and minimal resource use. Using recycled materials and solar-powered pumps, some systems further reduce environmental impact. Aquaponics provides urban communities with fresh produce, reduces the carbon footprint associated with food transport, and allows residents to grow their own food sustainably.

6.4.6 Case Studies of Circular Agriculture Eco-Innovations in Vietnam

6.4.6.1 Agroforestry in the Northern Highlands

In mountainous areas like Ha Giang and Son La, agroforestry has become a valuable circular agriculture practice. Farmers grow a mix of crops, including tea, fruit trees, and high-value spices like cinnamon and star anise, alongside native tree species. This integrated approach helps control soil erosion, enriches soil health through natural composting from fallen leaves, and supports biodiversity. Farmers use organic compost and biochar produced from crop residues, reducing chemical fertilizer use and enhancing soil fertility. The inclusion of trees also captures carbon, making agroforestry an effective strategy for climate resilience in the region. Agroforestry systems help stabilize hillsides, prevent landslides, and allow farmers to diversify their income through the sale of fruits, spices, and timber, contributing to both ecological and economic sustainability.

6.4.6.2 Tea Waste Recycling in Thai Nguyen Province

Thai Nguyen province, famous for its tea production, faces challenges with the disposal of tea waste, which can become a pollutant if not properly managed. Local farmers have started recycling tea waste into organic compost and natural insect repellents. Some tea farms are adopting vermicomposting, where tea waste is mixed with earthworms to create nutrient-rich organic fertilizer. This eco-innovation not only repurposes waste but also improves soil quality, benefiting subsequent tea crops. By turning tea waste into fertilizer, farmers reduce dependency on chemical inputs, save on disposal costs, and create a circular system where waste supports new growth. Additionally, the composted tea waste has led to a noticeable improvement in soil health and crop quality.

6.4.6.3 Duck-Rice Farming Systems in the Red River Delta

In the Red River Delta, particularly in provinces like Nam Dinh and Hai Duong, duck-rice farming is a traditional practice that has gained renewed interest as a circular agriculture model. Farmers raise ducks in rice paddies, where the ducks naturally control pests by eating insects and weeds, and their droppings fertilize the rice. Farmers have started integrating duck-rice farming with organic practices, using natural fertilizers and minimal pesticides to create a balanced ecosystem. They are also experimenting with new duck breeds better suited for rice paddies, which improve pest control efficiency. This system reduces the need for chemical pest control and fertilizers, leading to cost savings for farmers. The sale of ducks adds an income stream, and consumers benefit from both high-quality rice and duck meat, boosting local economies.

6.4.6.4 Livestock Manure Composting for Mushroom Farming in Bac Giang Province

In Bac Giang, livestock manure has been a valuable resource for mushroom farming, which is becoming increasingly popular in northern Vietnam. Local farmers compost livestock manure and use it as a nutrient source for growing mushrooms like oyster and shiitake, creating a closed-loop system where waste from one process supports another. The composting process uses biodigesters that convert manure into methane for biogas, used by farmers as a clean energy source. The residual compost, rich in organic nutrients, is then used in mushroom cultivation.

This closed-loop system reduces waste and provides farmers with both a renewable energy source and an additional income from mushroom sales. Additionally, composting manure mitigates methane emissions, which would otherwise contribute to greenhouse gases.

6.4.6.5 Intercropping with Medicinal Herbs in Lao Cai and Yen Bai Province

In the mountainous areas of Lao Cai and Yen Bai, intercropping medicinal herbs, such as turmeric and ginger, with staple crops like maize and cassava is gaining popularity. These herbs grow well in the shade of staple crops, creating a mutually beneficial system. By incorporating medicinal herbs that naturally repel pests and diseases, farmers reduce the need for chemical pesticides. Additionally, farmers are experimenting with locally adapted organic pest control, like neem extracts, to protect both staple crops and herbs. This practice not only provides farmers with a diversified income but also improves soil health and helps control pests naturally. The medicinal herbs add value to local agriculture, tapping into growing demand for herbal medicine in both domestic and international markets.

6.4.6.6 Organic Vegetable Production with Biofertilizers in Hanoi's Suburbs

Near Hanoi, smallholder farmers are increasingly turning to organic vegetable production, spurred by consumer demand for safe, chemical-free produce. Many of these farmers are adopting biofertilizers made from agricultural waste, such as rice husks and vegetable residues, to enrich the soil. Local producers are using microorganisms like Azotobacter and Phosphate-solubilizing bacteria in their biofertilizers, which improve nutrient availability in the soil, enhancing vegetable growth and yield without the need for synthetic fertilizers. This approach supports soil fertility, reduces input costs, and meets the growing demand for organic vegetables in urban markets. The biofertilizer innovation also promotes a circular system, as farmers recycle their waste into valuable agricultural inputs.

6.4.6.7 Composting, Biogas Production and Biofertilizers

Composting has gained traction in Vietnamese agriculture as a way to recycle agricultural waste and produce organic fertilizers. All wastes in agricultural production such as manure, urine of cattle, poultry, straw, and others are processed to become resources. This model has been implemented by agricultural producers in different localities across the country. The model includes "biogas technology" and "composting". In the biogas technology model, people employed biogas facilities to process processed (pig manure, chicken manure, cattle dung) into gas to use as domestic fuel. For composing strategy, by-products from cultivation (straw, corn, beans), domestic waste, and livestock waste (pig manure, chicken manure, cattle dung) are decomposed as organic fertilizer. Biofertilizers, which contain beneficial microorganisms, are also increasingly used to enhance nutrient availability and support plant growth. Modern composting systems incorporate microbial inoculants and controlled aerobic composting processes, which accelerate decomposition and improve compost quality. Additionally, biochar, a carbon-rich fertilizer made from organic waste, is being used to enhance soil structure and water retention. Composting reduces the need for chemical fertilizers, which can lead to soil and water contamination. Biofertilizers and biochar improve soil health, reduce input costs, and promote a circular system where agricultural waste is returned to the land as valuable nutrients.

Biogas production is an important eco-innovation in Vietnam, especially in areas with high livestock populations, such as the Mekong Delta and northern highlands. In this system, livestock manure is processed in biogas digesters, which capture methane emissions and convert them into biogas for cooking, heating, or even electricity generation. The byproduct of this process, a nutrient-rich slurry, is used as an organic fertilizer. Recent advances include small-scale, portable biodigesters that are accessible to smallholder farmers. Some systems are equipped with improved efficiency in methane capture, making biogas production cleaner and more effective. Biogas reduces greenhouse gas emissions from manure decomposition, decreases reliance on fossil fuels, and lowers household energy costs. The use of nutrient-rich slurry also enhances soil fertility, contributing to healthier crop yields.

6.4.6.8 Silvo-Pastoral Systems in Hoa Binh Province

In Hoa Binh province, farmers are implementing silvo-pastoral systems, where livestock graze in areas with trees. This approach merges forestry with animal husbandry, with trees providing shade and windbreaks for livestock while also generating income from timber and fruit. Some farmers use rotational grazing practices and plant nitrogen-fixing trees like acacia to improve soil quality and manage pasture health. The integration of animals and trees promotes a balanced ecosystem, where animals help fertilize the soil, and trees improve carbon capture. Silvo-pastoral systems prevent land degradation, provide diversified income from multiple sources such as timber, livestock, fruit, and build climate resilience. This model has been effective in restoring degraded lands and maintaining productivity in a sustainable, circular fashion.

6.4.6.9 Zero-Waste Farms

At a more advanced level, surpassing the models aimed at reducing, recycling, and establishing ecological connections, the application of circular economy principles extends to the industrial-scale agricultural production carried out by corporations. Notable examples include the biosecurity livestock model, known as 4Fs model (Farm-Food-Feed-Fertilizer) by Quế Lâm, production model by the T&T159, "green circle" model by the Vietnam Dairy Corporation (Vinamilk), and the circular model by the TH True Milk Group. These models are designed holistically based on the principle of zero waste in agricultural production. They operate under the premise that the output of one production process serves as input for another, facilitated by the application of advanced science and technology. 4F (Farm-Food-Feed-Fertilizer) model is considered as the most comprehensive circular economy in agriculture which has been initiated by Que Lam Group in 2020. 4F is a closed cycle which includes pig raising, annual crop production, animal feed and microbial fertilizer manufacturing. The waste from pig raising is collected and processed into microbial organic fertilizers for cultivation. Implementing the 4F model helps to increase economic efficiency, prevent disease, contribute to ecological environmental protection, and reduce greenhouse gas emissions. So far, the 4F model has been scaled up to 30 provinces in Vietnam. The manure that is processed by microbiological techniques is provided to farmers who grow maize, watermelon, tea, and dragonfruit across the country. In the 4F, the CE principles are applied from agricultural production to consumption (Ninh et al., 2023).

6.5 Technological Advances in Circular Agriculture

In recent years, the integration of technology into circular agriculture practices is helping to enhance traditional methods, optimize resource use, improve agricultural sustainability and make circular agriculture more efficient and scalable in Vietnam.

6.5.1 Precision Agriculture and Digital Tools

Digital technology is transforming Vietnamese agriculture by enabling farmers to manage resources more effectively. Digital transformation in agriculture is the process of changing the traditional agricultural production model to the agricultural model that uses technical technology as the foundation, applying technology to all agricultural production activities, maximizing raw materials and waste by-products to recycle into products or input materials, creating a circular mechanism in production, contributing to environmental protection, forming a circular and sustainable economic model in the future (Anbumozhi, 2022). Precision agriculture tools, such as drones, satellite imagery, and soil sensors, allow farmers to monitor crop health, manage water use, and apply fertilizers and pesticides in a targeted manner. Digital platforms also connect farmers to real-time data on weather, market prices, and sustainable farming practices.

Innovations include mobile platforms that allow farmers to receive market updates and technical advice, as well as sensor systems that automate irrigation based on real-time soil moisture levels. Precision agriculture reduces resource waste, improves crop yields, and enables farmers to respond to environmental changes quickly. Digital platforms make circular practices more accessible by providing farmers with the information needed to make informed decisions (Quy, 2024). Automated feeding systems for livestock help reduce food waste by delivering the exact amount of feed based on animal size, type, and activity. This minimizes feed costs and enhances livestock productivity. Automated feeders and aerators for fish ponds, controlled by mobile apps or sensors, help maintain optimal water quality and feeding schedules in integrated rice-fish farming, boosting fish productivity while reducing labor. Drones equipped with cameras and sensors help farmers monitor crop health, detect pests, and assess nutrient needs. This data-driven approach allows for precise application of fertilizers and pesticides, reducing chemical inputs and enhancing crop health in a sustainable way. Sensors are now used to monitor soil moisture, weather, and water levels in rice paddies and fields, enabling precise, real-time water management. This technology conserves water and reduces waste by ensuring that crops receive the exact amount needed.

6.5.2 Biogas Production and Waste-to-Energy Systems

Vietnamese livestock farms, particularly pig and poultry farms, have a tradition of using biogas systems. Manure from livestock is fed into biogas digesters, where it decomposes anaerobically to produce methane gas. This gas is then used as a cooking fuel or an energy source, reducing the reliance on fossil fuels. The byproduct of biogas production, a nutrient-rich slurry, can be used as an organic fertilizer for crops, completing the nutrient cycle. Modern biogas systems are now more efficient and affordable, making them accessible to smallholder farmers. New biogas technology allows for greater methane capture, increasing energy output and reducing greenhouse gas emissions. Innovations in mobile biogas systems also enable farmers in remote areas to adopt biogas solutions, reducing their dependence on non-renewable energy sources and contributing to circular nutrient cycles. Traditional bio-digesters are now enhanced with digital controls and real-time monitoring to improve efficiency. These systems convert livestock waste into biogas for cooking and heating, while the nutrient-rich residue is used as organic fertilizer. Large-scale farms sometimes implement biogas plants that produce electricity from animal waste, providing a renewable energy source for the farm and even nearby communities.

6.5.3 Aquaculture Wastewater Recycling

Aquaculture is a major industry in Vietnam, especially in the Mekong Delta. Advances in wastewater treatment and recycling enable fish farms to treat and reuse water within the farming system. This reduces water consumption, minimizes the release of pollutants into natural water bodies, and helps maintain healthier fish stocks. Circular aquaculture systems that treat wastewater and repurpose nutrient-rich byproducts for crop irrigation or animal feed enhance the sustainability of Vietnam's aquaculture sector.

6.5.4 Biochar Technology for Soil Improvement and Renewable Energy

Biochar is a charcoal-like substance created from organic waste through pyrolysis. It is rich in carbon and can be added to soil to enhance fertility and water retention. In Vietnam, crop residues like rice husks and corn cobs can be converted into biochar, which is then applied to fields. This technology not only improves soil health but also captures carbon in the soil, mitigating greenhouse gas emissions and promoting sustainable land management.

Rice straw and rice husk are carbonized by a farm household heater in a very low oxygen or sometimes no concentration to become biochar. The heater can be operated very easily without adding any energy but making the biochar quantity and quality stable. The heater then is re-designed with some improvement in size and construction material that can be used for a group of farmers on a village scale. Biochar then continues to be produced from many other crop residues like corn cobs, corn stems and leaves, bamboo and coconut residues. Biochar is used not only for soil improvement, but also for treatment of polluted water, making pot soil, tooth path and other forms of medicines. The char process is then optimized by using gasification system to gasify crop residues (rice straw, rice husk, saw dust) for cooking. The residues from that process are biochars remained in the gasifier for soil improvement material (Trinh, 2023).

6.5.5 Aquaponics and Vertical Farming

Combining aquaculture and hydroponics, aquaponic systems allow Vietnamese farmers to grow fish and vegetables together in controlled environments. Fish waste provides nutrients for plants, and plants naturally filter the water, creating a sustainable, closed-loop system. For high-value crops, some farms are using vertical farming systems with LED lighting and controlled environments. This technology optimizes space, reduces water use, and allows for year-round production.

Urban aquaponics is a growing trend in cities like Ho Chi Minh City and Hanoi, where small-scale systems enable residents to grow fish and vegetables in a closed-loop system. In aquaponics, fish waste provides nutrients for plants, which in turn filter the water for the fish. This system reduces water use and recycles nutrients efficiently, making it ideal for urban environments where space and resources are limited. Compact aquaponics systems use recycled materials and solar-powered pumps to minimize environmental impact. Some setups even use vertical farming techniques to maximize space efficiency. Aquaponics provides fresh, locally grown food, reducing the carbon footprint associated with food transport. It also empowers urban communities to contribute to food security and practice sustainable agriculture in their own spaces.

6.6 Synergies Between Traditional Practices and Technology

Synergies between traditional practices and modern technology in circular agriculture create innovative pathways for sustainable food production and resource management (Praveen et al., 2024). Traditional agricultural methods, such as crop rotation, composting, and mixed farming, emphasize natural cycles and resource efficiency. When combined with modern technologies like precision farming, remote sensing, and bioengineering, these practices can be scaled and optimized. For instance, traditional composting techniques can be enhanced with anaerobic digesters to convert organic waste into bioenergy and nutrient-rich fertilizer. Similarly, indigenous knowledge of pest control can work alongside advanced data-driven pest monitoring systems to minimize chemical usage. This fusion not only maximizes resource efficiency and reduces waste but also promotes ecological balance and resilience, fostering a more sustainable and regenerative agricultural system.

6.6.1 *Enhancing Traditional Techniques with Data-Driven Insights*

By combining traditional farming practices with data-driven precision agriculture, Vietnamese farmers can optimize inputs and improve crop yields. For example, traditional crop rotation practices can be informed by data on soil nutrient levels and pest dynamics, enabling farmers to select the best crop combinations and planting schedules.

6.6.2 *Improving Biogas Systems with Modern Technology*

Traditional biogas systems can be upgraded with modern technologies to increase efficiency and reach. Mobile biogas units, for instance, allow small-scale livestock farmers in remote areas to benefit from biogas energy. Moreover, advanced biogas digesters that capture more methane and reduce emissions align well with both traditional waste recycling practices and modern environmental goals.

6.6.3 *Integrating Biochar into Crop Residue Management*

Traditional practices of burning rice straw and crop residues can be replaced by biochar production, a practice that adds value and captures carbon. By producing biochar from agricultural waste, farmers can improve soil health while reducing greenhouse gas emissions, offering a sustainable alternative to crop residue burning.

6.6.4 *Using IoT for Monitoring Traditional Rice-Fish Systems*

IoT devices can help monitor water quality, temperature, and oxygen levels in traditional rice-fish systems, optimizing the environment for both fish and rice. These systems can increase productivity, improve fish survival rates, and reduce resource consumption, enhancing the circular benefits of this traditional practice.

6.7 Challenges and Solutions

6.7.1 Challenges in Implementing Eco-Innovations in Vietnam's Agricultural Sector

6.7.1.1 Policy and Regulatory Gaps

Government policies play a crucial role in fostering circular agriculture. While there is growing policy support for sustainable agriculture, there are still gaps in regulatory frameworks, infrastructure, and subsidies that could accelerate the adoption of eco-innovations. There is still a lack of specific incentives and regulatory frameworks to promote circular practices. The implementation of legal documents in this area is still ineffective, and there is a need for complete standards and technical regulations on products, goods, and waste to support the application of circular economy measure. Creating regulatory frameworks that promote circular practices, such as biogas incentives or organic certification subsidies, can drive adoption. Policies that encourage technology transfer, partnerships, and investment in rural infrastructure will further support the circular agriculture transition (Tram, 2020).

6.7.1.2 Limited Access to Technology and Infrastructure

While circular agriculture has immense potential, the technology required for advanced recycling, waste processing, and biogas production is often costly and not widely available, particularly for smallholder farmers. Limited access to infrastructure and technology impedes large-scale adoption. Smallholder farmers in Vietnam often lack the financial resources to invest in advanced technology, which can limit the adoption of circular agriculture practices. Setting up systems such as biogas digesters or digital precision tools requires capital, which can be prohibitive for small-scale farmers (Tram, 2020).

6.7.1.3 Knowledge and Training Gaps

Many farmers in Vietnam rely on traditional practices and lack awareness of circular agriculture benefits. Transitioning requires training in composting, waste management, and integrated systems, which can be a barrier in rural areas with limited educational resources. Insufficient training on the benefits and applications of eco-innovations makes it difficult for farmers to adopt circular practices effectively. There is insufficient awareness among stakeholders, including businesses, enterprises, and farmers, about the role, benefits, nature, content, and criteria of circular agriculture (Tram, 2020).

6.7.1.4 Limited Motivation Among Major Agricultural Producers and Businesses

The lack of pressure and incentives for major agricultural producers and businesses hinders their transition to a circular economy model. The agricultural value chain does not currently have well-established links where circular economy practices can be applied, and the connections between producers and businesses are still weak, preventing the closure of the circular chain (Ninh et al., 2023).

6.7.2 Recommennded Solutions for Enhancing Eco-Innovation in Circular Agriculture

6.7.2.1 Develop and Implement Policy Frameworks

National and regional policies should incentivize sustainable practices, promote recycling of agricultural waste, and provide the regulatory support needed to make eco-innovations accessible to farmers. Study found that the factors affecting the adoption of circular economy practices are Government policies, Production scale, Business Awareness, Technology, and Financial Investment (Nguyen et al., 2024). Introduce policies mandating sustainable agricultural practices, supporting research on eco-innovations, and providing subsidies for organic fertilizers and biofertilizers. National policies should promote sustainable practices, encourage recycling of agricultural waste, and incentivize circular agriculture at the local level. These policies can also address regulatory barriers that may impede eco-innovation adoption. The government should enact supportive policies to assist citizens and businesses in transitioning from current production models to circular agriculture practices. Support can include financial grants and technical advisory services (Chung et al., 2024).

A roadmap should be developed to guide the implementation of circular agriculture, starting from the micro level and gradually scaling up to the macro level. Agricultural producers and businesses should be the central driving force in this process. The roadmap should also incorporate financial mechanisms to support the achievement of circular agriculture goals in agriculture. Developing support mechanisms and policies (finance, preferential loans, land, technology) to attract business investment in building circular agriculture and the participation of the private sector in research and development activities, transfer and application of science and technology on circular agriculture (Chung et al., 2024; Ninh et al., 2023).

6.7.2.2 Education and Training Programs

Government and development agencies should invest in training programs to educate farmers on the benefits and applications of eco-innovations. These programs can also facilitate knowledge sharing among communities, building local expertise in circular practices. Programs can be hosted at local agricultural hubs to make information accessible. Both the central government and local authorities should enhance awareness among citizens about the long-term benefits of circular agriculture practices. Communication efforts should focus on raising awareness among producers about environmental protection issues and the benefits of circular agriculture. Producers, traders, and consumers should be encouraged to participate in collection and recycling activities. Support in terms of capital and technology should be provided, especially for the development of markets for secondary materials and recycled products. Public awareness about circular agriculture should be raised through targeted communication strategies and training programs at various educational levels, including high schools, colleges, universities, and short-term training courses. The aim is to change perceptions regarding production and consumption, promoting the use of renewable energy, fully recyclable and reusable products, and emphasizing resource recovery from used product (Chung et al., 2024).

6.7.2.3 Provide Financial Support for Eco-Innovations

Subsidies, low-interest loans, and grants should be made available to help farmers invest in biogas systems, precision farming tools, and other eco-innovations. Policies offering subsidies, grants, or low-interest loans would support farmers in investing in biogas systems, precision agriculture tools, and other eco-innovations. Investments in technology development and application in agriculture play a significant role in Circular Economy adoption. State budgets can facilitate technology development while reducing costs for businesses and households (Nguyen et al., 2024). Increasing access to affordable technology is essential for smallholder farmers. Partnerships with private companies and international organizations can facilitate investments in low-cost biogas digesters, composting systems, and water treatment facilities. Microfinancing and cooperative models can also help make these technologies accessible. Financial incentives, such as tax breaks and low-interest loans, would support farmers' adoption of costly eco-innovations, including biogas systems and precision farming tools (Chung et al., 2024).

6.7.2.4 Encourage Public-Private Partnerships

Partnerships between government, private sector, and development agencies can enhance access to eco-innovations by funding research, creating pilot programs, and supporting sustainable business models. Collaboration between government, private companies, and development agencies can enhance access to eco-innovations, fund pilot projects, and provide support for sustainable business models. The partnerships between government bodies, development agencies, research institutions, and farmer cooperatives could promote best practices in circular agriculture (Chung et al., 2024). Establish government and NGO-funded programs that provide eco-innovations at subsidized rates for smallholder farmers.

6.7.2.5 Invest in Digital Infrastructure

Digital infrastructure investment plays a crucial role in advancing circular agriculture by enabling resource efficiency, data-driven decision-making, and sustainable practices. Robust digital tools, such as IoT devices, sensors, blockchain systems, and AI-powered platforms, can monitor and optimize every stage of agricultural production, from soil health to waste recycling. Investments in reliable internet connectivity, cloud computing, and mobile platforms empower farmers to adopt precision farming techniques, trace product lifecycles, and integrate circular systems like composting, nutrient cycling, and renewable energy use. By leveraging such infrastructure, circular agriculture not only reduces waste and environmental impact but also enhances productivity and resilience in farming communities. These technologies create interconnected ecosystems that ensure transparency, sustainability, and scalability, paving the way for a regenerative agricultural economy (Anbumozhi, 2022).

6.8 Conclusions

Vietnam's agricultural sector stands at a critical juncture where sustainable, circular models are essential for balancing productivity with environmental health. Circular agriculture, powered by eco-innovations, offers an effective solution by optimizing resource use, reducing waste, and building climate resilience. By committing to circular practices, Vietnam can improve resource efficiency, reduce environmental impacts, and secure its agricultural sector's future in the face of climate challenges. With a mix of traditional knowledge and innovative practices, Vietnam is well-positioned to become a regional leader in circular agriculture, providing a model for other countries aiming to transition toward sustainable food systems. By investing in eco-innovations and creating supportive frameworks, targeted investments in technology and training, Vietnam can lead in circular agriculture, contributing to a resilient agricultural sector that ensures food security, environmental health, and economic prosperity for future generations. This approach not only aligns with Vietnam's goals for sustainable development but also enhances food security, rural livelihoods, and resilience to climate change. As Vietnam moves toward a circular agricultural future, it sets an example for other countries in Southeast Asia and beyond, showcasing the power of combining tradition and innovation to achieve sustainability.

References

Anbumozhi, V. (2022). Role of innovation and digital transformation to promote the circular economy. In Institute of Strategy, Policy on Natural Resources and Environment, *Promoting innovation toward actualizing the circular economy in Viet Nam*. Institute of Strategy, Policy on Natural Resources and Environment.

Chung, D. K., Duy, L. V., & Loan, L. T. T. (2024). Circular agriculture: A general review of theories, practices, and policy recommendations. *Vietnam Journal of Agricultural Sciences, 7*(2), 2173–2184.

Ellen MacArthur Foundation. (2016). Cities and circular economy for food. https://www.ellenmacarthurfoundation.org/publications/cities-andcircular-economy-for-food

Helgason, K. S., Iversen, K., & Julca, A. (2021). *Circular agriculture for sustainable rural development: The environmental impact of conventional agriculture* (Policy Brief No 105). Department of Economic and Social Affairs, United Nations.

MONRE, 2020. Third biennial update report to the united nations framework convention on climate change. Ministry of Natural Resources and Environment, Hanoi.

Nguyen, K. T., Khue, N. T. M., & Tran, Q. P. (2024). Examining the factors influencing the level of circular economy adoption in agriculture: Insights from Vietnam. *Research on World Agricultural Economy, 5*(1), 48–58. https://doi.org/10.36956/rwae.v5i1.992

Ninh, N. T. H., Le, N. P., & Chung, D. K. (2023). How to enhance circular agriculture models at micro-level in Vietnam? A review. *Current World Environment, 18*(2), 483–493.

Praveen, K., Abinandan, S., Venkateswarlu, K., & Megharaj, M. (2024). Synergy of eco-innovation with on-farm practices enhances circularity beyond conventional nutrient recovery framework. *Resources, Conservation and Recycling, 208*(2024), 107735.

Quy, N. H. (2024). The role of digital transformation in the circular economy in Vietnam. *Journal of State Management, 31*(11), 7–16.

Selvan, T., Panmei, L., Murasing, K. K., Guleria, V., Ramesh, K. R., Bhardwaj, D. R., et al. (2023). Circular economy in agriculture: Unleashing the potential of integrated organic farming for food security and sustainable development. *Frontiers in Sustainable Food System, 7*, 1170380.

Son, H. N., & Tuan, B. T. (2022). Mu Cang Chai Rice Terraces and its water management, The cultural heritages of water in Eastern and South-Eastern Asia. *ICOMOS Thematic Study*, 151–155.

Son, H. N., Chi, D. T. L., & Kingsbury, A. (2019). Indigenous knowledge and climate change adaptation of ethnic minorities in the mountainous regions of Vietnam: A case study of Dao people in Bac kan province. *Agricultural Systems, 176*(2019), 102683. https://doi.org/10.1016/j.agsy.2019.102683

Son, H. N., Ha, H., & Kingsbury, A. (2021). Indigenous knowledge and the enhancement of community resilience to climate change in the Northern Mountainous Region of Vietnam. *Agroecology and Sustainable Food Systems, 45*(4), 499–522. https://doi.org/10.1080/21683565.2020.1829777

Tram, P. T. (2020). Circular economy in agriculture in Vietnam and solutions for development. *Journal of Environment, 2020*, 9(2020).

Trinh, M. V. (2023). Circular economy in Vietnam's agriculture, the potential substitute of chemical fertilizers by animal wastes and crop residues. *FFTC Journal of Agricultural Policy, 4*, 1–13.

World Bank. (2022). *Spearheading Vietnam's green agricultural transformation: Moving to low-carbon rice*. World Bank.

Chapter 7
Eco-Innovation in Jammu & Kashmir: Catalyzing Sustainable Development Through Circular Economy Practices

Amita Gupta

Abstract This chapter explores the role of **eco-innovation** and **circular economy** practices in promoting **sustainable development** in **Jammu & Kashmir (J&K)**. The region, known for its diverse natural resources and unique socio-economic challenges, faces significant environmental issues such as **resource depletion**, **waste management**, and **energy insecurity**. The circular economy offers a comprehensive framework to address these challenges by emphasizing **resource efficiency**, **waste reduction**, and the **regeneration of natural systems**. Through **eco-innovation**, J&K has the potential to transform key sectors such as **waste management**, **renewable energy**, **agriculture**, and **eco-tourism** to create new **economic opportunities** while preserving the environment. The chapter highlights several **successful case studies** of eco-innovation in J&K, such as **waste-to-energy projects**, **solar energy** initiatives, and **sustainable agriculture practices**, demonstrating their potential to enhance **economic resilience**, **reduce environmental degradation**, and improve **public health**. The adoption of eco-innovation will not only address the region's immediate environmental concerns but also foster long-term economic growth, improve public health, and empower local communities.

Keywords Sustainable development · Clean production · Resource efficiency · Waste management · Renewable energy · Sustainable agriculture

---

A. Gupta (✉)
Centre for Comparative Religion and Civilization, Central University of Jammu, Rahyaa Suchani, Samba, Jammu and Kashmir, India
e-mail: amita.ccrc@cujammu.ac.in

S. Patnaik et al. (eds.), *Circular Economy and Sustainable Value Creation through Eco-Innovation*, Circular Economy and Sustainability,
https://doi.org/10.1007/978-3-031-88516-7_7

7.1 Introduction

Jammu & Kashmir, located in the northernmost part of India, is a region of significant **ecological**, **cultural**, and **economic** importance. Its diverse landscapes, from the majestic **Himalayas** to the lush, fertile valleys of Kashmir, are home to a wealth of **natural resources** and **unique biodiversity**. Traditionally, the economy of J&K has been based on **agriculture**, **handicrafts**, and **tourism**. However, these sectors are increasingly under pressure from **environmental degradation**, **climate change**, and socio-economic challenges. The region's reliance on **non-renewable resources**, coupled with **rapid urbanization** and **over-exploitation** of natural resources, has led to rising **waste generation**, **pollution**, and **resource depletion**.

In response to these mounting challenges, **eco-innovation** and **circular economy principles** have gained prominence as vital strategies for promoting **sustainable development** in J&K. **Eco-innovation**, which focuses on creating new products, services, and processes that contribute to environmental sustainability, offers a promising solution to mitigate the region's environmental impact. The **circular economy**, centered on **reduce, reuse, and recycle** model which presents a transformative framework for rethinking resource management, minimizing waste, and optimizing efficiency across sectors.

The need for circular economy principles in J&K is particularly urgent due to the region's **fragile ecosystems**, which are susceptible to the effects of **climate change**, including **glacier melt**, **water scarcity**, and extreme weather events. Traditional practices such as **deforestation** for fuelwood and **unsustainable water use** in agriculture exacerbate these issues, while increasing urban populations and tourism further strain **waste management systems** and contribute to pollution.

This chapter explores the potential of eco-innovation and circular economy practices to address these pressing challenges. By integrating these practices into **renewable energy**, **agriculture**, **waste management**, and **eco-tourism**, J&K can reduce its reliance on non-renewable resources, promote clean energy, and ensure long-term environmental and economic resilience. Through case studies and policy analysis, the chapter highlights opportunities for scaling these solutions and overcoming the barriers to their widespread adoption.

7.2 Conceptualizing Eco-Innovation and Circular Economy in Jammu and Kashmir

7.2.1 Defining Eco-Innovation

Eco-innovation refers to the development of new products, processes, or services that provide **environmental benefits** while also contributing to **economic** and **social** value. Unlike traditional innovations, which focus solely on economic gains, eco-innovation integrates **sustainability** into the design, production, and consumption phases. In **Jammu & Kashmir (J&K)**, eco-innovation offers a solution to the region's challenges of **environmental degradation**, **resource scarcity**, and **socio-economic development** while simultaneously creating new avenues for growth and environmental protection.

7.2.2 Key Characteristics and Principles of Eco-Innovation

Eco-innovation is driven by several key principles that can be applied to various sectors within J&K:

- **Resource Efficiency**: This principle emphasizes optimizing resource use, reducing consumption, and increasing output. In J&K, sectors like **agriculture** and **forestry** can benefit from **precision farming** and **sustainable forestry management**, helping conserve natural resources and minimize environmental impact.
- **Waste Minimization**: Central to eco-innovation is the reduction of waste generation and the repurposing or recycling of waste. J&K, with its growing urbanization and tourism, faces significant waste management challenges. **Waste-to-energy technologies** offer solutions by converting organic waste into usable energy, addressing both waste disposal and energy needs.
- **Sustainable Design**: Sustainable design focuses on creating products and processes that have minimal environmental impact. In J&K, this could include **green buildings**, **eco-friendly construction materials**, and sustainable practices in the **handicraft sector** (e.g., using natural dyes and biodegradable materials).
- **Clean Technologies**: Technologies like **solar power**, **wind energy**, and **bioenergy systems** are central to J&K's transition toward sustainability. By adopting these technologies, the region can reduce its dependence on **fossil fuels**, mitigate **pollution**, and enhance **energy security**.

7.2.3 Relationship Between Eco-Innovation and Circular Economy

Eco-innovation and **circular economy principles** are closely linked. While eco-innovation focuses on developing environmentally friendly technologies and services, the circular economy extends these innovations by promoting **resource recovery**, **product life extension**, and **regeneration of natural systems**. The **circular economy** aims to eliminate the traditional "take, make, dispose" model, encouraging instead the continuous reuse and recycling of resources.

In J&K, circular economy practices can complement eco-innovation by ensuring that **waste materials** are reused or recycled, reducing the need for new raw materials and decreasing environmental impact. For example, in **agriculture**, **organic farming**, **composting**, and **water conservation technologies** are key circular economy strategies that enhance sustainability while conserving resources.

7.2.4 Circular Economy Framework

The circular economy (CE) framework focuses on maximizing resource use through the principles of **reduce**, **reuse**, and **recycle**, which not only reduce environmental harm but can also stimulate economic growth.

- **Reduce**: Emphasizing resource minimization, this principle applies to industries such as **agriculture**, where sustainable farming practices reduce chemical use, and in buildings, where **energy-efficient technologies** lower resource consumption.
- **Reuse**: The reuse principle involves extending the product and material lifecycle. In J&K, this could be applied in the **handicrafts** and **construction sectors**, where materials like wood and stone are reused rather than discarded. **Water reuse** through **rainwater harvesting** and wastewater treatment can conserve this precious resource.
- **Recycle**: Recycling reduces the need for virgin materials and reintroduces them into the production cycle. In J&K, recycling programs for **plastic** waste and agricultural by-products can reduce waste while creating resources for new products like **biogas** or **compost**.

7.2.5 Triple Bottom Line: Environmental, Social, and Economic Sustainability

The **Triple Bottom Line (TBL)** approach assesses sustainability in three areas: **environmental**, **social**, and **economic**.

- **Environmental Sustainability**: Circular economy principles help reduce environmental footprints by conserving resources and promoting clean technologies. In J&K, this could involve **waste-to-energy** initiatives, **sustainable land management**, and **water conservation**.
- **Social Sustainability**: Eco-innovation fosters **community involvement**, creating **green jobs** in **renewable energy**, **eco-tourism**, and **sustainable agriculture**. Social sustainability also involves improving public health through reduced pollution and fostering equity and inclusivity.
- **Economic Sustainability**: Circular economy practices reduce operational costs, promote **innovation**, and open new markets for green products. In J&K, sectors like **eco-tourism**, **green construction**, and **organic farming** offer opportunities for **sustainable economic growth** while preserving natural resources.

Eco-innovation and circular economy principles provide a promising path to address J&K's environmental, social, and economic challenges. The region can achieve resource efficiency, environmental conservation, and economic resilience by incorporating these practices into key sectors. Overcoming barriers such as **infrastructure constraints**, **financial limitations**, and **awareness** will require coordinated efforts from state government, private sectors and local communities. With the right support, J&K can transition to a sustainable, circular economy that benefits both people and the environment.

7.3 The Role of Eco-Innovation in Sustainable Development in Jammu and Kashmir

Eco-innovation is crucial for **sustainable development** as it integrates environmental sustainability with **economic growth** and **social well-being**. In **Jammu & Kashmir (J&K)**, eco-innovation offers a transformative approach to address the region's unique environmental and socio-economic challenges, ensuring long-term sustainability across key sectors like **resource management**, **renewable energy**, and **agriculture**.

7.3.1 Sustainable Resource Management

Efficient resource management is vital for J&K's sustainability, as the region is rich in **water**, **soil**, and **forests** yet faces pressures from **over-extraction** and **climate change**. Eco-innovation can optimize resource use, ensuring their availability for future generations while minimizing environmental degradation.

Precision farming and **smart irrigation systems** can improve water efficiency, reduce chemical use, and increase agricultural yields. Technologies like **drip irrigation**, which delivers water directly to plant roots, can significantly reduce water consumption, a crucial factor in a region facing water scarcity. **Rainwater harvesting** systems can further help store water for agricultural use during dry periods.

In addition, **waste-to-energy technologies** can address growing waste generation, especially in urban areas. Converting organic waste into **biogas** or **electricity** can reduce landfill waste while providing a renewable energy source, contributing to both **waste management** and **energy security**.

7.3.2 Promoting Renewable Energy Solutions

J&K's dependence on fossil fuels for energy is unsustainable, and transitioning to **renewable energy** is critical. The region has ample potential for **solar**, **hydropower**, and **biomass** energy.

Solar energy, particularly in **Ladakh**, where sunlight is abundant, offers a promising solution. **Small hydropower projects** harnessing the region's rivers can also provide clean electricity, especially in remote villages. However, scaling these solutions in J&K's rugged terrain is challenging due to **grid connectivity** issues. **Off-grid solar systems**, **micro-hydro projects**, and **biogas plants** can provide sustainable energy to isolated communities, reducing reliance on conventional fuels (Pradhan & Jha, 2021). **Bisma et al. (2020)** highlight how the adoption of **alternative fuels** in the **foothills of the Western Himalayas** has significantly reduced **fuelwood consumption**, offering a model for sustainable energy transition in the region. In 2023, the **Jammu & Kashmir Energy Development Agency (JAKEDA)** launched the **Solar Panel Subsidy in Jammu & Kashmir** program to promote the use of **renewable energy sources** in the region (JAKEDA, 2023).

7.3.3 Sustainable Agriculture and Eco-Innovation

Agriculture in J&K is underpinned by traditional practices, often relying on **chemical fertilizers** and **water-intensive irrigation**. Transitioning to **sustainable agriculture** is essential for reducing environmental degradation and improving food security. Alai (2023) highlights the significance of integrating innovation and sustainability in the horticultural sector, suggesting that eco-friendly practices such as integrated pest management and organic farming can significantly improve soil health and reduce dependency on harmful chemicals. In J&K, organic farming can reduce chemical inputs, enhance soil fertility, and open access to higher-value markets for organic produce.

Organic farming can reduce chemical inputs and enhance soil fertility. J&K's **Kashmir Valley** has seen success in this regard, with farmers gaining access to higher-value markets for organic produce. Additionally, **waste-to-resource agriculture**, which uses agricultural waste for **composting** or **biogas production**, reduces waste while creating valuable resources for farming.

Water conservation is particularly critical in J&K, where agriculture accounts for the largest portion of water consumption. **Water-efficient technologies** such as **drip irrigation** and **rainwater harvesting** can help improve water use efficiency, reduce pressure on local water resources, and increase crop yields. These practices are vital in a region where **water scarcity** is a growing concern.

Eco-innovation is essential for J&K's sustainable development. The region can enhance environmental resilience, economic growth, and social well-being by embracing resource-efficient practices, renewable energy solutions, and sustainable agriculture. Though challenges remain, particularly in **scaling** these innovations, their long-term benefits—such as improved resource management, reduced environmental impact, and new economic opportunities—will be crucial for building a sustainable and equitable future for J&K.

7.4 Key Sectors for Eco-Innovation in Jammu & Kashmir

Jammu & Kashmir (J&K) is well-positioned to benefit from eco-innovation, particularly in sectors that are critical to its economic structure and environmental sustainability. These sectors include **waste management**, **green technology adoption**, and **sustainable tourism development**. By embracing eco-innovation, J&K can foster long-term sustainability, enhance local economies, and reduce its environmental footprint. This section explores the key sectors in the region where eco-innovation can make a significant impact, and how circular economy practices can be effectively incorporated into these industries to promote sustainability.

Waste Management and Recycling Initiatives: Jammu & Kashmir faces growing waste management challenges, especially in urban areas and popular tourist destinations. As the region grapples with increasing waste generation due to urbanization and tourism, **Waste-to-Energy Technologies** offer a promising solution. By converting **organic waste** into **biogas** or **electricity**, these technologies can help reduce landfill waste while simultaneously generating clean energy. This approach not only mitigates waste disposal challenges but also supports the region's energy security. Bisma et al. (2020) emphasize the importance of transitioning to alternative fuels, which can be further complemented by waste-to-energy technologies that address both waste management and energy production. Integrating such technologies aligns with the circular economy framework, which promotes **resource efficiency, reuse**, and **recycling**.

Circular Economy Practices in Urban and Rural Waste Management: Circular economy practices, such as waste segregation, recycling, and composting, can offer effective solutions to improve waste management across both urban and rural areas. In cities like Srinagar and Jammu, where waste production is rapidly increasing due to urban growth and tourism, comprehensive waste management programs are essential. Implementing a waste segregation system that separates organic, recyclable, and non-recyclable materials right at the source can reduce the amount of waste sent to landfills and improve recycling rates.

In rural areas, where agricultural waste is abundant, **waste-to-energy technologies**, such as biogas production, can help turn this waste into renewable energy and compost. These technologies not only address waste management but also provide a cleaner, more sustainable energy source, reducing reliance on firewood and helping to mitigate the environmental impact of deforestation. By integrating these circular economy practices, J&K can not only manage waste more effectively but also promote cleaner energy, benefiting both the environment and local communities.

Success Stories and Challenges in Waste-to-Energy Projects: Several **waste-to-energy projects** have been initiated in J&K, with mixed results. One such project is the **Srinagar Waste-to-Energy Plant**, which has had initial success in processing organic waste and converting it into biogas. However, as noted by Kumar (2023), scaling up these initiatives faces challenges such as **insufficient infrastructure**, **high initial costs**, and **lack of public awareness**. To overcome these barriers, the government must provide **financial incentives**, **policy support**, and **public-private partnerships (PPPs)** to promote waste-to-energy systems across the region.

Green Technology Adoption: Jammu & Kashmir is ripe for adopting **green technologies** in its key industries, including **agriculture**, **handicrafts**, and **energy production**. Green technologies not only contribute to environmental sustainability but also promote **economic growth** by creating new industries, reducing operational costs, and improving efficiency.

Clean Production Technologies for Local Industries: The region's **handicrafts** and **agriculture** sectors, which are deeply ingrained in the local economy, stand to benefit significantly from adopting clean production technologies. For example, in the **handicrafts sector**, using eco-friendly materials and sustainable production methods can drastically reduce environmental harm. By incorporating natural dyes and biodegradable materials, local artisans can create products that are not only environmentally friendly but also appealing to both domestic and international markets, especially as demand for sustainable goods continues to rise.

In **agriculture**, adopting clean production technologies like **organic farming**, **precision agriculture**, and **biopesticides** can reduce reliance on harmful chemicals, improve soil health, and boost productivity. Technologies such as **GPS-based systems** and **soil sensors** enable farmers to optimize the use of water, fertilizers, and pesticides, leading to higher yields and more sustainable farming practices. These innovations also enhance resilience to unpredictable weather patterns, which is especially crucial in a region like J&K that is increasingly impacted by **climate change** (Sreedhar & Rao, 2019).

7.5 The Role of Innovation in Traditional Sectors like Handicrafts, Agriculture, and Tourism

Jammu & Kashmir's traditional industries—**agriculture**, **handicrafts**, and **tourism**—stand to benefit from the integration of green technologies and sustainable practices. **Eco-tourism**, for example, offers a way to mitigate the environmental impacts of traditional tourism while supporting sustainability. Jammu & Kashmir's tourism industry, which has long been a key economic driver, can also benefit from eco-innovation. Idrees et al. (2020) emphasize that by focusing on sustainable practices within the tourism sector, J&K can not only boost its economy but also ensure that the region's fragile ecosystems are protected. By incorporating green technologies into tourism development, J&K can maintain its cultural heritage and natural beauty while attracting eco-conscious travelers. By promoting eco-friendly travel options, reducing waste, and protecting natural landscapes, J&K can enhance its tourism sector and attract eco-conscious visitors, helping to preserve the region's cultural and natural heritage for future generations. The development of **eco-friendly accommodations**, **waste reduction strategies**, and **sustainable tourism practices** can help preserve J&K's natural landscapes and cultural heritage while providing economic opportunities for local communities (Bashir et al., 2020).

In the context of sustainable innovation, green intellectual capital has become an important factor in fostering both environmental and social innovations. According to Sharma et al. (2022), the integration of green intellectual capital with social innovation plays a significant role in driving sustainable practices, particularly in sectors like agriculture, tourism, and handicrafts. This approach can enable Jammu & Kashmir to unlock the full potential of eco-innovation by combining knowledge, creativity, and environmental awareness to build resilient, sustainable communities. Similarly, **green technologies** in agriculture, such as **rainwater harvesting**, **solar-powered irrigation systems**, and **water-efficient farming techniques**, can help make farming more sustainable and water-conserving. **Agro-tourism** could also thrive in the region, blending agriculture with tourism in a way that simultaneously promotes environmental conservation and economic growth.

7.5.1 Sustainable Tourism Development

Tourism is one of the most important sectors in Jammu & Kashmir, contributing significantly to the local economy. However, the growing number of tourists has placed significant pressure on the region's natural resources and infrastructure. **Eco-tourism** presents a sustainable alternative that can help mitigate the environmental impact of mass tourism while offering economic benefits to local communities. According to Idrees et al. (2020), the tourism sector in Jammu & Kashmir has the potential to drive both economic growth and environmental sustainability, provided that sustainable tourism practices are adopted to minimize the negative impacts on the region's natural resources. In regions like Jammu & Kashmir, eco-tourism offers a promising solution to the challenges of over-tourism and environmental degradation. Parray et al. (2021) suggested that integrating eco-friendly practices into tourism hotspots like Gulmarg and Pahalgam can help preserve the region's natural heritage while offering economic benefits to local communities.

7.5.2 Eco-Tourism and Waste Management Strategies for Tourism Hotspots

Tourism hotspots like **Katra**, **Gulmarg**, **Pahalgam**, and **Sonmarg** are facing challenges such as **waste accumulation**, **pollution**, and **over-tourism**. Implementing eco-friendly waste management systems, including waste segregation and recycling, can help these areas manage tourist-generated waste more effectively. Additionally, **rainwater harvesting**, **composting**, and **biogas production** systems can be introduced in these tourist areas to improve sustainability.

Eco-tourism can also promote sustainable practices among tourists, such as reducing plastic waste, conserving water, and supporting local communities. Eco-tourism can play a significant role in mitigating the environmental impact of mass tourism by implementing eco-friendly practices such as waste segregation and water conservation. Studies like those of Bashir et al. (2020) suggest that eco-tourism not only helps preserve the region's natural landscapes but also supports economic growth by attracting eco-conscious travelers.

7.5.3 Circular Economy Approaches for the Hospitality Industry

The **hospitality industry** in J&K, including hotels, resorts, and restaurants, has a significant environmental impact. The hospitality sector can reduce its ecological footprint by adopting circular economy approaches, such as waste reduction, energy-efficient systems, and sustainable sourcing of food and materials. **Solar panels**, **rainwater harvesting systems**, and **eco-friendly building materials** can be incorporated into the design of hotels and resorts to make them more sustainable.

Moreover, **zero-waste initiatives**, **food waste recycling**, and **energy-efficient appliances** can reduce costs and improve the environmental performance of the hospitality sector. These efforts not only contribute to **sustainability** but also enhance the appeal of J&K as a **green tourism destination**.

The key sectors for eco-innovation in Jammu & Kashmir—**waste management**, **green technology adoption**, and **sustainable tourism**—offer significant opportunities for advancing **sustainability** and **economic growth** in the region. By adopting circular economy principles and **eco-innovation** strategies, J&K can reduce its environmental footprint, create new economic opportunities, and foster **social equity**. Scaling up these efforts will require **policy support**, **financial investment**, and **collaborative partnerships** between the public and private sectors, as well as active involvement from local communities. Through the successful integration of eco-innovation into these sectors, J&K can lead the way in **sustainable development** in the Himalayan region and beyond.

7.6 Barriers and Challenges to Eco-Innovation in Jammu & Kashmir

While eco-innovation offers significant promise for sustainable development in Jammu & Kashmir (J&K), several barriers impede its widespread adoption. These include infrastructure limitations, financial constraints, cultural resistance, and, importantly, socio-political factors that influence the successful implementation of eco-innovation. Addressing these obstacles is crucial for unlocking the region's potential and achieving its sustainable development goals.

7.6.1 Infrastructure Limitations

A major barrier to eco-innovation in J&K is its underdeveloped infrastructure, particularly in areas such as waste management, renewable energy generation, and sustainable agriculture.

- **Waste Management:** In cities like Srinagar and Jammu, waste management systems are outdated and insufficient. Recycling facilities are limited, and waste segregation systems are not widespread, impeding the ability to recycle or repurpose waste. This leads to excessive waste accumulation, environmental degradation, and dependence on inefficient resource management systems.

Renewable Energy Infrastructure: J&K has vast potential for solar, hydropower, and biomass energy, the necessary infrastructure to support these technologies is lacking. Many remote areas remain disconnected from the grid, complicating efforts to deploy solar or small hydropower projects effectively. Even in areas with grid access, the infrastructure is often outdated and ill-equipped to accommodate renewable energy inputs. Scaling up green technologies, such as precision farming tools or solar panels, requires substantial investment in both infrastructure and technical expertise—resources that many local stakeholders, particularly farmers and small businesses, lack

7.6.2 Financial and Institutional Constraints

Financial limitations represent another significant barrier to the widespread adoption of eco-innovation in J&K. Many small and medium-sized enterprises (SMEs), farmers, and local governments face difficulties in accessing the capital required for green technologies.

- **Access to Capital:** The lack of green financing options—such as subsidies, low-interest loans, or grants—makes it difficult for stakeholders to transition to sustainable practices. Renewable energy projects, such as solar power systems and small hydropower plants, require large capital investments that many local businesses and rural communities cannot afford. Furthermore, financial institutions in J&K are often hesitant to fund eco-innovation projects due to perceived high risks and a lack of understanding of their long-term benefits (Kumar, 2023).
- **Government Policy:** While supportive policies are crucial, J&K suffers from policy inconsistencies and bureaucratic inefficiencies. Inter-departmental coordination is often lacking, and policies related to eco-innovation are not always aligned with circular economy principles. There is also a lack of long-term policy consistency, which undermines investor confidence. Strengthening policy frameworks and introducing financial incentives, such as tax breaks and subsidies for green projects, is essential to overcome these financial barriers.

7.6.3 Cultural and Social Barriers

Cultural and social resistance to change is another significant challenge to eco-innovation in J&K. The region's traditional industries, particularly agriculture and handicrafts, are deeply embedded in local practices, which can slow the transition to more sustainable methods.

- **Resistance in Traditional Sectors:** Many farmers continue to rely on chemical fertilizers, pesticides, and water-intensive irrigation methods despite the growing evidence supporting the benefits of organic farming and precision agriculture. Similarly, the handicraft sector, which is a major part of the local economy, often adheres to resource-intensive production methods that resist change.
- **Public Awareness:** There is also a general lack of public awareness regarding circular economy principles and the benefits of eco-innovation. Many rural communities may not fully understand the advantages of resource efficiency, waste reduction, or sustainability. As a result, there is limited interest in adopting these practices, especially when initial costs are perceived as high or when the long-term benefits are not immediately apparent.

7.6.4 Socio-Political Barriers

While infrastructure and financial constraints are important, the socio-political factors in Jammu & Kashmir play a pivotal role in either facilitating or hindering the adoption of eco-innovation. The region's political instability, governance challenges, and complex socio-political dynamics significantly impact the pace and scale of eco-innovation initiatives.

- **Governance and Political Instability:** J&K's political landscape, marked by frequent changes in governance and policy shifts, creates an uncertain environment for eco-innovation projects. Political instability can lead to inconsistent policy support for sustainable development initiatives, making it difficult for stakeholders to plan and invest in long-term eco-innovation projects. A lack of continuity in leadership and priorities also results in slow implementation of environmental policies, reducing investor confidence in eco-innovation. Sheikh (2023) outlines the significant structural changes in Jammu and Kashmir's economy, particularly in light of political instability and shifting governance structures. These changes have a direct impact on the region's ability to adopt eco-innovation practices, as the political climate often dictates the prioritization of sustainability initiatives and the availability of resources for green technology deployment.
- **Local Political Dynamics:** The local political landscape, characterized by differing priorities among political parties and stakeholders, influences the adoption of eco-innovation. Policies promoting sustainability may not always align with local political agendas, which can lead to resistance or inadequate support for eco-innovation initiatives. Moreover, political power struggles may also affect the equitable distribution of resources required for scaling eco-innovation solutions.
- **Community and Institutional Coordination:** Effective eco-innovation requires strong collaboration between local communities, governmental bodies, NGOs, and businesses. However, there are often gaps in coordination and trust among

these stakeholders. Local governance structures, affected by political fragmentation, may lack the capacity to effectively support circular economy initiatives. Moreover, the involvement of local communities in decision-making processes is often limited, which hinders the adoption of sustainable practices.

The **barriers to eco-innovation** in Jammu & Kashmir are multifaceted and require a **holistic approach** to overcome. **Infrastructure limitations, financial constraints, cultural resistance,** and **socio-political factors** must all be addressed to enable the widespread adoption of circular economy practices. By improving governance, aligning policies with **sustainable development goals**, providing **financial incentives**, and fostering **community engagement**, J&K can unlock the potential of eco-innovation and pave the way for a more **sustainable and resilient future**.

7.7 Policy and Institutional Frameworks for Promoting Circular Economy

The successful adoption of eco-innovation and circular economy (CE) practices in Jammu & Kashmir (J&K) hinges on the development of **robust government policies**, **effective financial mechanisms**, and strong **public-private collaborations**. This section **delves** into the existing government initiatives, the role of public-private partnerships (PPPs), and outlines **strategic policy recommendations** to overcome infrastructure and financial constraints, thereby scaling up circular economy practices in the region.

7.7.1 Government Initiatives and Support

J&K has taken initial steps toward supporting environmental sustainability, but a more **integrated** and **expansive policy framework** is needed to facilitate the widespread adoption of circular economy principles.

- **Waste Management**: The **Municipal Solid Waste Management Rules 2016** and the **Swachh Bharat Mission** are pivotal in promoting waste segregation and recycling. However, these initiatives face significant barriers, such as **inadequate infrastructure** and **limited public awareness**. To address these challenges, the incorporation of **circular waste management practices**, including **waste-to-energy technologies**, should be prioritized to enhance waste recovery, reduce landfill reliance, and generate renewable energy.
- **Renewable Energy**: J&K has substantial potential for **renewable energy generation**, particularly from **solar** and **hydropower sources**. Existing policies such as the **J&K Solar Policy** (2013) and the **Small Hydro Power Policy** pro-

vide a strong foundation for renewable energy projects. However, scaling these projects, especially in remote and off-grid areas, remains a major challenge due to **poor grid connectivity** and **infrastructure limitations**. To address this, policies should promote the integration of **energy storage solutions**, **waste-to-energy systems**, and **microgrids**, which would not only enhance energy access but also contribute to reducing reliance on fossil fuels.

- **Sustainability Initiatives**: The **J&K Economic Reconstruction Plan** touches upon sustainability but lacks a comprehensive approach that connects diverse sectors, including waste management, energy, agriculture, and tourism. A **holistic policy approach** that links these sectors under a circular economy framework is crucial for promoting sustainable growth and resilience in J&K. This approach should focus on **resource efficiency**, **waste minimization**, and the maximization of recycling across sectors.
- **Irrigation Infrastructure and Pollution Control**: The **Ranbir Canal**, built by the Dogra ruler Maharaja Ranbir Singh in the late nineteenth century, stands today as an exemplary piece of engineering, originally designed to irrigate the city of Jammu and the surrounding areas like **Ranbir Singh Pura**. However, over the years, the canal channels have become heavily polluted by waste, largely due to local communities discarding garbage into the waterway. Despite the canal's historical importance, the government has failed to take sufficient action to prevent this pollution or implement effective waste management practices along the canal. Addressing this issue is critical, as the canal remains an important resource for irrigation and water supply, and its contamination poses both environmental and public health risks. A circular economy approach, including waste collection systems and eco-friendly disposal mechanisms, could be integrated to restore and maintain the canal's ecological balance (Gupta et al., 2024).

7.7.2 Role of State Governments

State governments play a pivotal role in promoting the transition to a circular economy but are often hindered by **financial constraints**, **bureaucratic inefficiencies**, and lack of technical capacity. Despite these challenges, local authorities can spearhead eco-innovation initiatives by adopting the following measures:

- **Incentivizing Sustainable Practices**: The government can offer **tax incentives** or **grants** to encourage investments in **green infrastructure**, such as recycling centers, biogas plants, and green construction technologies. This can stimulate private sector participation and foster the growth of eco-innovative businesses.
- **Public-Private Partnerships (PPPs)**: **PPPs** are an essential mechanism to bridge the financial and technical gaps in scaling eco-innovation projects. By leveraging both **public resources** and **private sector expertise**, PPPs can help accelerate the implementation of circular economy practices.

7.7.3 Public-Private Partnerships (PPP) in Eco-Innovation

PPPs play a critical role in scaling **green infrastructure** and **eco-innovative projects** in J&K. Through effective collaboration between the public and private sectors, these partnerships can help overcome the **financial** and **technical barriers** that often impede the adoption of sustainable practices.

- **Waste-to-Energy**: Collaborations between the government and private companies in the waste-to-energy sector can enable the **conversion of organic waste into biogas or electricity**. This would not only reduce the region's dependence on fossil fuels but also promote **renewable energy generation**, thereby contributing to the region's **energy security** and sustainability goals.
- **Renewable Energy**: J&K's abundant **solar** and **hydropower resources** present significant opportunities for PPPs. Private sector investment in **solar mini-grids** and **small hydropower projects** can bring electricity to off-grid rural areas, with the government providing **financial incentives**, **policy support**, and **regulatory frameworks** to ensure the long-term sustainability of these initiatives.
- **Eco-Tourism**: The tourism sector in J&K can greatly benefit from PPPs that promote **eco-friendly tourism infrastructure**. Collaborative efforts can help develop **sustainable tourism practices** that preserve the region's natural beauty while simultaneously creating **economic opportunities** for local communities.

7.7.4 Case Studies of Successful Collaborations

Successful examples of PPPs from other regions can provide valuable insights into the potential for similar collaborations in J&K. Notable case studies include:

- **Bangalore Waste-to-Energy Project**: This project demonstrates the potential for PPPs to overcome challenges in waste management and energy generation. By leveraging private sector expertise and public resources, the project has succeeded in **converting waste into biogas and electricity**, reducing landfill use and generating renewable energy.
- **Ladakh Solar Energy Projects**: In Ladakh, PPP models have been employed to establish **solar energy systems** that provide electricity to off-grid villages. These collaborations have proven effective in overcoming **technical** and **financial barriers**, ensuring energy access in remote areas.

These case studies highlight the effectiveness of **PPPs** in scaling eco-innovation projects and overcoming infrastructure and financial challenges.

7.7.5 *Policy Recommendations for Scaling Circular Economy Practices*

To scale circular economy practices in J&K, the following policy recommendations are crucial:

1. **Green Tax Incentives**: Introduce **tax breaks** and financial incentives for businesses and individuals investing in **eco-friendly technologies**, such as renewable energy systems (solar, wind) and **energy-efficient building practices**. These incentives will reduce the upfront costs of transitioning to greener alternatives.
2. **Green Financing Mechanisms**: Establish dedicated **green finance funds** to provide **low-interest loans**, **grants**, or **subsidies** for projects in the **renewable energy**, **waste management**, and **eco-tourism sectors**. Such financing mechanisms would lower the **financial risk** for private investors and make it easier for eco-innovative startups to access capital.
3. **Extended Producer Responsibility (EPR)**: Enforce regulations that require manufacturers to take responsibility for the lifecycle of their products, including **recycling** and **disposal**. This will encourage the design of products that are more **durable**, **recyclable**, and **energy-efficient**, thus reducing the overall environmental footprint.
4. **Circular Economy Policies**: Develop comprehensive, sector-wide policies that integrate **circular economy principles** into **waste management**, **energy production**, **agriculture**, **construction**, and **tourism**. These policies should prioritize **resource efficiency**, **waste minimization**, and **sustainable practices**, ensuring that all sectors work in synergy to reduce environmental impact.
5. **Capacity Building and Education**: Invest in **training** and **capacity-building programs** to equip J&K's workforce with the skills needed for the **green economy**. These programs should focus on emerging technologies, sustainable practices, and **green entrepreneurship** to prepare the next generation of **eco-innovators** and leaders.
6. **Inclusive Development for Marginalized Communities**: J&K's migrant worker population, primarily from states like **Madhya Pradesh**, **Uttar Pradesh**, and **Bihar**, often lives in precarious conditions. **Inclusive development policies** should prioritize the provision of **sustainable livelihoods**, **education**, and **skill development** for these workers, allowing them to transition to **green jobs** in sectors like renewable energy, waste management, and eco-tourism. **Training centers** and **job opportunities** in these areas will improve their living conditions while fostering **inclusive economic growth**.

The transition to a circular economy in J&K requires a **well-coordinated policy framework**, **innovative financial mechanisms**, and **collaborative efforts** between the government, private sector, and local communities. By implementing the recommendations outlined in this section, J&K can create an **enabling environment** for eco-innovation, drive **sustainable economic growth**, and achieve **long-term**

environmental sustainability. Through **targeted policy interventions**, **green finance**, and robust **public-private partnerships**, J&K has the potential to become a **model of sustainable development** in the Himalayan region and beyond, leading the way in **circular economy practices** for other states and regions to follow.

7.8 Case Studies of Eco-Innovation in Jammu & Kashmir

Jammu & Kashmir (J&K) has made notable progress in adopting eco-innovation to address its environmental and socio-economic challenges. This section highlights key examples of **waste-to-energy**, **renewable energy**, and **sustainable agriculture** initiatives in the region, demonstrating the potential of these innovations to drive a sustainable future.

7.8.1 Waste-to-Energy Projects

Waste-to-energy (WTE) technologies provide dual benefits—managing waste and generating clean energy. In J&K, waste management is a growing concern due to urbanization, tourism, and agricultural waste.

- **Rural Biogas Projects**: In rural areas, biogas plants convert **agricultural residues**, **animal manure**, and **household waste** into **renewable energy** and **fertilizer**. A successful example is the biogas program in the **Kashmir Valley**, where such initiatives reduce reliance on firewood and kerosene, provide clean cooking energy, and produce organic compost for sustainable farming (Singh & Thakur, 2018). Despite their success, scaling these projects requires technical support, investment, and local awareness.
- **Municipal Waste-to-Energy in Srinagar**: In Srinagar, a waste-to-energy plant converts **municipal solid waste (MSW)** into **biogas** and **electricity**, addressing waste disposal challenges in the city. However, this initiative faces barriers, including inefficient waste segregation, inadequate infrastructure, and public resistance. Greater community engagement and improved waste management practices are essential for success.

7.8.2 Renewable Energy Implementation

J&K's diverse geography offers significant potential for **renewable energy** generation, particularly in **solar power** and **small hydropower** projects.

- **Solar Power in Ladakh**: The **Ladakh Solar Power Project** has deployed **solar photovoltaic systems (PVs)** in off-grid villages, reducing the reliance on kero-

sene. Solar-powered **microgrids** have improved energy access and quality of life in remote areas. Despite these successes, challenges remain in scaling solar infrastructure and ensuring long-term maintainability in harsh climatic conditions. Public-private partnerships (PPPs) are critical to expanding these projects.
- **Small Hydropower in Jammu**: The **Madhopur Power Project** in Jammu utilizes **small hydropower** to generate clean electricity for local communities. These small-scale projects are particularly effective in remote areas with limited grid connectivity. However, challenges like **land acquisition**, **environmental concerns**, and **financing** hinder their expansion. Policy support, including **subsidies** and **grants**, is necessary to accelerate these projects.

7.8.3 Sustainable Agriculture Practices

Agriculture is the backbone of J&K's economy, yet traditional practices have led to **soil degradation**, **water scarcity**, and pollution. Eco-innovation offers sustainable solutions to enhance productivity while protecting the environment.

- **Organic Farming in Kashmir**: The **Kashmir Organic Farmers Association (KOFA)** promotes **organic farming** to reduce chemical inputs, restore soil fertility, and improve sustainability. Organic farming has led to increased yields and access to niche markets, particularly for **exported organic products**. However, transitioning to organic methods requires investment in **training**, **market linkages**, and **organic inputs**.
- **Precision Agriculture and Water Conservation in Jammu**: In Jammu, **precision agriculture** technologies like **drip irrigation**, **soil moisture sensors**, and **weather forecasting** tools optimize water use and improve crop yields. These systems are particularly valuable in water-scarce areas, as they reduce water consumption and enhance irrigation efficiency. Additionally, **rainwater harvesting** systems capture water during the monsoon for use in drier months, reducing irrigation costs and conserving water (Sreedhar & Rao, 2019).

The case studies presented show that eco-innovation in J&K—through **waste-to-energy**, **renewable energy**, and **sustainable agriculture**—has led to significant environmental, economic, and social benefits. However, scaling these initiatives faces challenges such as **infrastructure limitations**, **financial constraints**, and **public awareness**. Overcoming these barriers will require **policy support**, **financial incentives**, and **collaborative partnerships**. By addressing these challenges, J&K can unlock the full potential of eco-innovation and create a sustainable, resilient future for the region.

7.9 Opportunities for Circular Economy in Jammu & Kashmir

Jammu & Kashmir (J&K) stands at a crucial juncture in its development, with an opportunity to shift towards a circular economy (CE) that emphasizes resource efficiency, waste reduction, and the regeneration of natural systems. The region's abundant natural resources—such as water, sunlight, and biomass—coupled with its rich cultural heritage, provide fertile ground for adopting circular economy practices. However, to truly catalyze sustainable development, it is essential to integrate local community perspectives and challenges. The region's transition to a circular economy can be greatly influenced by the active participation of grassroots communities, who are often at the heart of resource use, waste generation, and the management of ecosystems. This section highlights the economic, social, and environmental benefits of adopting circular economy practices in J&K, with a focus on empowering local communities and overcoming grassroots challenges.

7.9.1 Economic Benefits of Eco-Innovation

Eco-innovation, driven by circular economy principles, can significantly boost J&K's economy by fostering new markets, reducing dependence on external resources, and improving resilience to environmental and economic shocks. For these benefits to be fully realized, however, grassroots challenges, such as limited access to financing, technology, and market information, must be addressed.

- **Green Job Creation**: Circular economy practices offer a wide array of **green jobs** in sectors such as renewable energy, waste management, sustainable agriculture, and eco-tourism. For instance, renewable energy projects—such as solar power, small hydropower, and biogas—demand skilled workers for installation, operation, and maintenance. Similarly, the waste-to-energy sector offers employment opportunities in waste collection, sorting, and processing. However, a major grassroots challenge lies in ensuring that **local communities**, especially in remote areas, have access to the necessary training, resources, and job opportunities. To bridge this gap, local vocational training programs should be established, focusing on **solar panel installation**, **biogas plant maintenance**, and **waste management**.
- **Local Economic Opportunities**: By supporting sustainable agricultural practices like **organic farming** and **precision agriculture**, J&K can tap into both local and global markets for eco-friendly products. These sectors can create new market linkages, benefiting **local farmers** and agribusinesses. For example, small-scale farmers can be incentivized to adopt organic farming practices that are aligned with circular economy principles, promoting the use of **natural fertilizers** and **composting**. To ensure these benefits reach the grassroots level, tar-

geted policy interventions and **financial incentives** are essential, particularly in rural areas where farmers face challenges such as access to capital, land tenure issues, and market connectivity.

7.9.2 Diversification of Local Economies

Traditionally, J&K's economy has relied heavily on agriculture, handicrafts, and tourism. However, **eco-innovation** can diversify the local economy and help J&K thrive in a **sustainable** manner by integrating circular economy practices into existing sectors.

- **Eco-Friendly Handicrafts**: The handicraft sector is integral to J&K's cultural identity and economy. By integrating **eco-innovation** into this sector, J&K can produce sustainable products using biodegradable materials, natural dyes, and recycled fibers. This could help create a niche market for **eco-friendly handicrafts**, attracting environmentally conscious consumers. Grassroots communities, particularly **artisan groups** in rural areas, could benefit from such initiatives, provided they receive access to training, design resources, and markets.
- **Sustainable Tourism**: Eco-tourism presents a significant opportunity to promote both economic benefits and the preservation of J&K's natural landscapes and cultural heritage. The region's renowned tourist destinations, such as **Gulmarg** and **Pahalgam**, can serve as models for sustainable tourism. Implementing sustainable tourism practices—such as waste reduction, water conservation, and ecosystem protection—will not only attract eco-conscious travelers but also create long-term economic opportunities for local communities. However, to ensure that grassroots communities benefit from eco-tourism, **community-based tourism** models should be encouraged. These models emphasize local participation in decision-making and the equitable distribution of tourism revenues, funding **community development projects** in healthcare, education, and infrastructure.

7.9.3 Social and Environmental Benefits

The transition to a circular economy offers significant **social and environmental benefits** for J&K, such as reducing environmental degradation, improving public health, and empowering local communities. These benefits can be magnified by addressing **local community needs** and **challenges**.

- **Reducing Environmental Degradation**: Circular economy practices such as recycling, composting, and waste-to-energy can help reduce **landfill waste**, minimize **pollution**, and conserve natural resources. In local communities, particularly in **urban slums** or **rural areas**, waste management is a major challenge, with limited infrastructure for waste segregation and recycling. Community-led initiatives, such as **waste segregation** at the household level, can provide a more sustainable solution, if supported by local authorities with adequate training and infrastructure. Moreover, grassroots campaigns to raise **awareness about the impact of waste** on local ecosystems can empower communities to take responsibility for their waste.

- **Improving Public Health**: Sustainable agricultural practices, including organic farming, reduce the use of harmful chemicals, which in turn improves soil health, biodiversity, and the quality of local food. This leads to **better public health** by decreasing exposure to pesticides and chemicals linked to diseases such as cancer and respiratory issues. In rural areas, **organic farming cooperatives** could be supported to provide both income and health benefits to local farmers. Additionally, **eco-construction** and **green infrastructure** in urban areas can help enhance air quality and reduce energy consumption, promoting healthier living conditions for **local communities.**
- **Empowering Local Communities**: Eco-innovation can empower local communities by involving them in sustainable practices such as waste management, renewable energy production, and eco-tourism. For example, **biogas plants** can be introduced in rural areas to reduce reliance on firewood, while **solar microgrids** can provide clean electricity to off-grid villages. Local communities in areas with limited access to the grid can thus benefit from **decentralized renewable energy** solutions. In addition, **eco-tourism** offers income opportunities, where **tourism revenues** can directly fund **community development projects** in healthcare, education, and infrastructure.

7.9.4 Educational and Skill Development Opportunities

A successful transition to a circular economy in J&K will depend on **a skilled workforce**. Training and education are essential for supporting eco-innovation and green technologies at the grassroots level.

- **Training the Workforce**: By offering specialized training in **renewable energy**, **eco-construction**, **waste management**, and **sustainable agriculture**, J&K can equip local workers with the necessary skills to support the circular economy. For example, **training programs for solar panel installation** and **biogas plant maintenance** can be designed to employ local youth and women in green sectors. Establishing local **skill development centers** that focus on **green technologies** will help ensure that local communities have the expertise needed to participate in eco-innovation projects.
- **Promoting Environmental Education**: Integrating **sustainability principles** into school and university curricula will foster a culture of environmental responsibility. This education will not only equip future generations with the tools to

drive circular economy practices but will also help to spread **awareness about the benefits of eco-innovation** among local communities. Grassroots education programs should be designed to engage communities in environmental conservation and waste management, ensuring that local populations actively contribute to sustainable development.

The opportunities for eco-innovation and circular economy practices in J&K are vast, offering **economic, social, and environmental benefits** that can transform the region's development. By focusing on **green job creation**, **sustainable industries**, **resource efficiency**, and **community empowerment**, J&K can build a resilient and inclusive economy. To fully realize these opportunities, strong support from **policies**, **financial incentives**, and **educational programs** is essential. **Community engagement** and **grassroots participation** will be crucial in overcoming local challenges and unlocking the full potential of a circular economy. With collaborative efforts from the government, the private sector, and local communities, J&K can embrace a **sustainable future** and become a model for **circular economy practices** in the Himalayas and beyond.

7.10 Conclusion

This chapter has explored the immense potential of **eco-innovation** and the **circular economy** as transformative tools for driving **sustainable development** in Jammu & Kashmir (J&K). Rich in natural resources yet grappling with significant **environmental**, **socio-economic**, and **infrastructural challenges**, J&K stands at a critical juncture where the adoption of circular economy principles can offer sustainable solutions. These principles, which emphasize **resource efficiency**, **waste reduction**, and the **regeneration of natural systems**, provide a path forward to address some of the region's most pressing issues, including **waste management**, **energy scarcity**, and **sustainable agriculture**.

Eco-innovation has the potential to **revolutionize key sectors** like **waste management**, **renewable energy**, **agriculture**, and **eco-tourism**, generating new economic opportunities while simultaneously preserving the environment. By embracing these practices, J&K can foster **green industries**, create **green jobs**, and reduce reliance on non-renewable resources, thus contributing to the improvement of **public health** and overall **quality of life** for its residents. In this process, **local communities** must be at the heart of the transformation, as they are both the primary stakeholders and the key to long-term success.

However, the path to implementing a circular economy is not without its challenges. **Infrastructure limitations**, **financial constraints**, and **cultural resistance** are some of the key barriers that could impede progress. To overcome these obstacles, a **holistic and inclusive approach** is essential—one that integrates the perspectives and needs of local communities while ensuring policy support, targeted financial mechanisms, and robust educational and capacity-building programs.

7.11 Future Outlook and Pathways Forward

The successful scaling of **eco-innovation** in J&K will require the expansion of existing pilot projects, such as **waste-to-energy systems**, **solar power initiatives**, and **sustainable agricultural practices**. Scaling these initiatives will necessitate significant investments in **infrastructure**, **technology**, and **local capacity building**. **Government policies**, including **subsidies**, **tax incentives**, and **green financing mechanisms**, will play a crucial role in supporting these ventures and ensuring that circular economy practices are integrated into regional development strategies.

7.11.1 The Role of Stakeholders

The transition to a circular economy in J&K requires **active participation from all stakeholders**, each of whom plays a unique and essential role in realizing the region's sustainable development goals:

- **Government**: The government must provide a **supportive policy framework**, incentivize **green infrastructure projects**, and create a **legal and regulatory environment** that encourages the adoption of circular economy practices. It is critical that circular economy principles are integrated into **regional development plans** to ensure long-term alignment with J&K's economic and environmental goals.
- **Private Sector**: The private sector is a key enabler of **eco-innovation**, with the capacity to drive investment in **green technologies** and create **public-private partnerships (PPPs)** for scaling **renewable energy projects**, **waste management systems**, and **eco-tourism infrastructure**. Businesses must be encouraged to invest in sustainable practices, both to reduce environmental impact and to tap into growing markets for green products and services.
- **Local Communities**: Local communities are the foundation of a successful circular economy. Engaging them in initiatives such as **waste management**, **renewable energy production**, and **sustainable farming** will ensure that these projects are **inclusive**, **equitable**, and **long-lasting**. Empowering communities with the knowledge, skills, and resources to implement sustainable practices is paramount. Grassroots participation in **eco-tourism**, **biogas production**, and **waste-to-energy systems** will create self-sustaining local economies, particularly in **rural areas** where infrastructure is often lacking.
- **Academic Institutions**: Universities and research centers play an indispensable role in the development and implementation of **eco-innovation**. By fostering

research and development in green technologies, promoting **skills training**, and offering **educational programs** in sustainability, these institutions will equip future leaders and entrepreneurs with the necessary expertise to drive J&K's circular economy forward.

7.11.2 A Transformative Shift Towards Sustainability

J&K is at a pivotal moment where **eco-innovation** and **circular economy practices** can drive a **transformative shift** towards sustainable development. The region's ability to transition into a green, sustainable future hinges on the active involvement of government bodies, businesses, local communities, and educational institutions. By embracing circular economy principles, J&K can unlock new economic opportunities, mitigate environmental challenges, and improve the well-being of its citizens.

The region's success in this transition will depend largely on **collaboration** and **partnerships** between these stakeholders. The development of **green industries**, the construction of **sustainable infrastructure**, and the promotion of **public-private partnerships** will collectively help build a **more resilient, inclusive, and sustainable economy**. With the right mix of **policy**, **financial support**, and **public engagement**, J&K has the potential to become a **model** for sustainable development in the Himalayan region and beyond.

7.11.3 Investing in a Greener Future

Investing in **eco-innovation**, **resource efficiency**, and **sustainable practices** will not only address immediate environmental concerns but will also ensure a **prosperous**, **environmentally secure**, and **socially equitable** future for the people of J&K. Through **commitment**, **innovation**, and **collaboration**, the region can **build a greener and more sustainable future** for all its residents, ensuring that the benefits of sustainable development are **felt at the grassroots level**.

References

Alai, S. (2023). Innovation and Sustainability Dynamics in the Horticultural Sector. Innovation and Sustainability Dynamics.

Bashir, U., Sharma, K., & Sharma, K. (2020). A comprehensive survey on ecotourism development in Jammu & Kashmir. *Journal of Sustainable Tourism, 29*(4), 559–576. https://doi.org/10.1080/09669582.2020.1746947

Bisma, J., Mushtaq, T., Gangoo, S. A., Sofi, P. A., Peerzada, I., & Malik, A. R. (2020). Impact of alternate fuels on domestic fuelwood consumption in the foothills of Western Himalayas. *Energy Policy, 136*, 111–118. https://doi.org/10.1016/j.enpol.2020.111118

Gupta, N., Khar, S., & Zaffar, O. (2024). Impact of Ranbir canal closure on wheat productivity in Jammu district. *Agricultural Engineering International: The CIGR e-journal, 26*(1), 42–49.

Idrees, M., Mir, A. A., & Tripathi, L. K. (2020). Economic and environmental development through the tourism sector in Jammu and Kashmir. *Tourism and Sustainability Review, 9*(4), 105–113. https://doi.org/10.1016/j.tour.2020.06.009

J&K Solar Power Policy (2013) *Jammu & Kashmir Government*.

Kumar, B. (2023). Impact of technological innovation for clean and green production on the environment in Jammu and Kashmir. *South Asian Journal of Social Studies and Economics, 17*(1), 1–6. https://doi.org/10.9734/sajsse/2023/v17i1626

Parray, P. A., Sharma, P., & Sharma, N. (2021). Eco-tourism: An influential impact on Jammu and Kashmir. *Environmental Management, 25*(3), 299–312. https://doi.org/10.1016/j.envman.2020.10.005

Pradhan, R., & Jha, A. (2021). Renewable energy transition in Jammu & Kashmir: A sustainable development perspective. *Renewable and Sustainable Energy Reviews, 147*, 111–118. https://doi.org/10.1016/j.rser.2021.111118

Sharma, V., Sheikh, D. A. R. B., & A.M. (2022). Green intellectual capital and social innovation: The nexus. *Journal of Intellectual Capital, 23*(6), 1199–1220. https://doi.org/10.1108/JIC-11-2020-0361

Sheikh, B.A. (2023) *Structural changes in Jammu and Kashmir economy*. Published 25-01-2023.

Singh, R., & Thakur, M. (2018). Waste-to-energy technologies for rural development in Jammu & Kashmir: Feasibility and challenges. *Energy Reports, 4*, 232–240. https://doi.org/10.1016/j.egyr.2018.05.001

Solar Panel Subsidy in J&K (2023) *Jammu & Kashmir Energy Development Agency (JAKEDA)*.

Sreedhar, A., & Rao, R. (2019). Sustainable agricultural practices and eco-innovation in the Himalayan region: A case study of Jammu & Kashmir. *Sustainability Science, 14*(3), 567–581. https://doi.org/10.1007/s11625-019-00793-w

Chapter 8
Circular Worth: A Critical Analysis of Recycled PET Price

Lambiase Nadia and Barbera Filippo

Abstract One of the key principles of the circular economy is to generate value from waste, that is, to undertake processes to enhance the value of matter so that it remains within production and consumption cycles as long as possible. This principle opens up several research questions: what does it mean to generate value? What is value and which is its relationship to price? To address these questions, we present a critical analysis of the recycled polyethylene terephthalate (r-PET)'s price and its value chain in Italy. Undertaking a price critical valuation means relying on an assessment of what the "real" price of the "thing" should be. In order to address this question, the price formation mechanisms and justification criteria for PET and r-PET provided by actors in the recycled PET supply chain have been explored. To test whether the price of r-PET, compared to the price of virgin PET, is able to carry the process of valuing the waste it incorporates, we explicitly focus on the social mechanisms that make sense of the respective prices of virgin and recycled PET and on the justifications given to these prices. The analysis has been grounded on the three key components of the commodity structure: qualification of objects, prices and value. The essay offers an insight about the concept of environmental cost to understand if and how these can help to build a better definition of asset valuation and valuation practices, in those case where monetary value is just one of the many possible forms of value attribution.

Keywords Circular economy · Commodification · Environmental cost · Constitutive incommensurability · Externalities · Price · Polyethylene terephthalate · Recycled polyethylene terephthalate · Recycling · Social cost · Valuation · Valuation practices · Value · Waste

L. Nadia (✉)
University of Turin, Turin, Italy
e-mail: nadia.lambiase@unito.it

B. Filippo
Collegio Carlo Alberto, Turin, Italy
e-mail: filippo.barbera@unito.it

S. Patnaik et al. (eds.), *Circular Economy and Sustainable Value Creation through Eco-Innovation*, Circular Economy and Sustainability,
https://doi.org/10.1007/978-3-031-88516-7_8

8.1 Introduction

One of the core principles of the circular economy is to generate value from waste (Lacy et al., 2016), that is, to undertake processes to enhance the value of matter so that it remains within production and consumption cycles for as long as possible. This principle opens up two key research questions: what is value and what is its relationship to price? Which mechanisms govern the transition from good to waste? To answer these questions, we present an empirical analysis of the price of recycled polyethylene terephthalate (r-PET) and its value chain in Italy.

The perspective from which we started in reconstructing the value chain of recycled PET in Italy is the one considered usual for the Italian context since 1997,[1] when the CONAI-COREPLA[2] consortium system was established by law. This perspective has been joined by a different one, which has emerged since the CORIPET[3] consortium came into play in 2016. In addition, we also chose to investigate the perspective of chemical recycling, given its growing interest, while knowing that to date, mechanical recycling is the prevailing methodology for the treatment of plastics and PET in particular (Ragaert et al., 2017). Finally, we found it worth to interview a nascent social enterprise that derives its business model from a variant of the philosophy and machinery of "Precious Plastic",[4] as a form of valorization of plastic material alternative to the classic chain of separate collection and subsequent recycling.

Once the value chain was reconstructed (Fig. 8.1), we proceeded to identify the variety of actors to be interviewed in order to capture the different viewpoints on the whole system, and in particular on price dynamics (r-PET and virgin PET), on the role of regulation and about the role the other actors played. In line with the ecological approach (Star, 1995), we have attempted to include and describe the different perspectives and different knowledge claims, making sure that the subjects' positions are visible and open to critical inquiry.

Figure 8.1 lists the actors interviewed, with their role/function within the value chain, the date the interviews were conducted, and the initial genesis of the contact.

The interviews took place between March and September 2021, via video call lasting about an hour and a half each. The structure of the interviews was the same for all actors involved, but depending on the actors some points were explored more

[1] Legislative Decree 22/1997, also known as the Ronchi Decree. Edoardo Ronchi is one of the interviewees.

[2] CONAI stands for Consorzio Nazionale Imballaggi, to which 6 supply chain consortiums belong, including COREPLA (National Consortium for the Collection, Recycling and Recovery of Plastic Packaging).

[3] CORIPET (Voluntary Consortium for PET Recycling) is an autonomous and specific consortium for only recycling of packaging made from PET polymer.

[4] The project's open source kit for artisanal plastic recycling is used, in most cases, within educational initiatives, several of which are part of research projects. In addition, 50 percent of the initiatives have a commercial profile, perhaps coupled with an educational and population engagement purpose (Spekkink W et al. 2020).

N.	COMPANY	ROLE	FOUNDATION	ITERVIEW DATE
1	COREPLA	NATIONAL CONSORTIUM-EPR	1997	26.03.2021 17.05.2021
2	CORIPET	AUTONOMOUS CONSORTIUM-EPR	2010 (operativo dal 2018)	18.03.2021 09.06.2021
3	Plastipak	PRODUCERS OF PACKAGING (PRE-SHAPE IN PET)	1967	-
4	Ferrarelle	PRODUCERS OF PACKAGING (BOTTLES)	1893	13.04.2021 23.04.2021
5	ROBOPLAST	PRODUCERS OF PACKAGING (NOT BOTTLES)	1979	23.12.2021 22.01.2021
6	ASSOSELE	SELECTORS ASSOCIATION	1996	03.08.2021
7	DEMAP S.R.L.	SELECTOR	2004	13.07.2021
8	CGR	MECHANICAL	2009	26.02.2021
9	ALIPLAST	MECHANICAL	1982	08.07.2021
10	DENTIS Recycling	MECHANICAL	1987	-
11	ASSORIMAP	MECHANICAL RECYCLERS	1978	06.04.2021
12	CARBO	CHEMICAL RECYCLER	1997	25.05.2021
13	GR3N	CHEMICAL RECYCLER	2011	18.05.2021
14		PRECIOUS PLASTIC	2021	18.05.2021
15	Unionplast	-	2005	-
16	PPR	INSTITUTE OF PROMOTION PLASTICS RECYCLING	2004	02.09.2021
17	FISE UNICIRCULAR	ASSOCIATION	2018	-
18	legacoopsociali	SOCIAL COOPERATIVE	1886	22.04.2021
19	CONFCOOPERATIVE	SOCIAL COOPERATIVE	1919	23.04.2021
20	TORINO METROPOLI	PUBLIC ADMINISTRATION	(1847)	11.06.2021
21	TORINO METROPOLI	PUBLIC ADMINISTRATION	(1847)	18.06.2021
22	a2a	MULTY UTILITY	2008	06.07.2021
23	plastic consult	CONSULTANT	1979	14.05.2021
24	materioteca plastic consult	PROJECT PLASTICS VALORIZATION	2008	07.07.2021
25	proplast	CONSULTANT	1998	25.05.2021
26	INTESA SANPAOLO INNOVATION CENTER	FINANCIAL SYSTEM	2014	02.07.2021
27	Edoardo Ronchi	EX PRIME MINISTER OF ENVIRONMENT	minister of environment	01.06.2021
28	UNI	ITALIAN STANDARDIZATION BODY	1921	07.09.2021

Fig. 8.1 Actors interview list

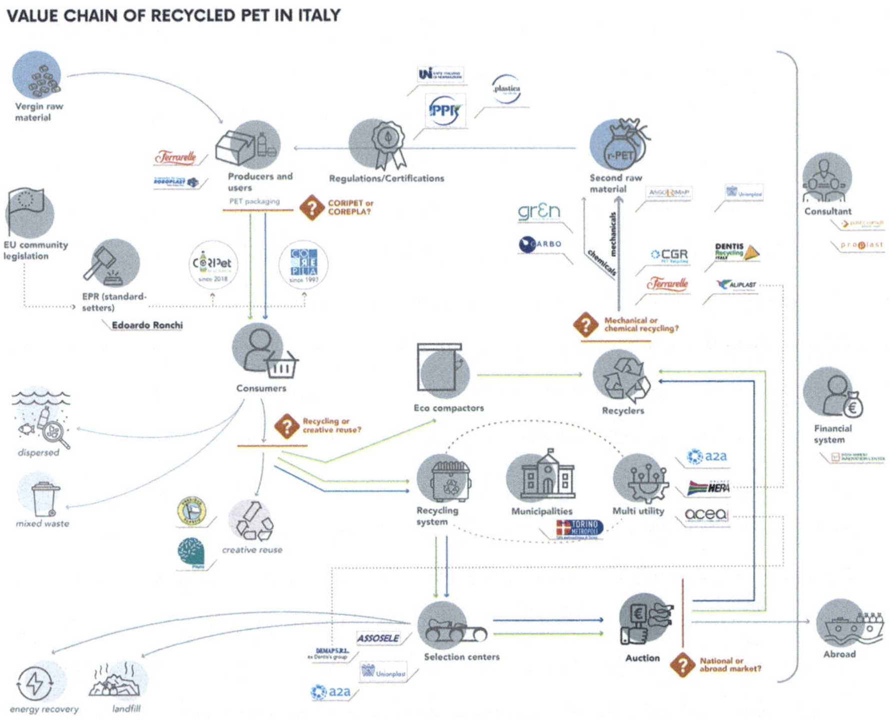

Fig. 8.2 Visual representation of the recycled PET value chain in Italy, and related actors interviewed

in detail, sometimes opening up new insights, while others were not addressed (Fig. 8.2).

For the purpose of the research question, to verify whether the price of r-PET, compared to the price of virgin PET, is able to adequately communicate the process of valorization of the waste it incorporates, we explicitly asked the interviewees, on the one hand, to detail the respective prices of virgin and recycled PET and, of the other, to justify these prices. To set up the development of the analysis, we chose to follow the triadic approach proposed by Boltansky and Esquerre (2019) related to the three components of the commodity structure: qualification of objects, prices and value.

Regarding the first point (qualification of objects) we have appealed to the normative corpus by highlighting how it, operating as an *agencement* (Calıskan & Callon, 2010) allows us to define what is waste and how it ceases to be waste, being able to acquire again the status of commodity. Regarding the second point (prices) we focus on the answers given by the interviewees regarding the explanation of prices for PET and r-PET, respectively. Finally, for the last item (value) we analyze whether and what appeals to value were fielded as a price justification device.

8.2 From Waste to Commodity: The Role of Legal Regulation

How can a material considered waste acquire the status of a commodity again? To answer this question, one must understand the institutional and normative instances that determine what a commodity is and give rise to market circulation. An undetermined thing cannot be valued: "If the thing is not determined, it is not possible to attribute to it a place in the universe of the commodity, to compare it with other things and thus to attribute to it a price with respect to other prices" (Boltansky & Esquerre, 2019 p.125). The law plays here a key role. The legal determination (as far as Europe is concerned) comes with Directive 75/442/EEC, replaced by Directive 2006/12/EC, then again replaced by the Waste Framework Directive 2008/98/EC in turn again amended in part by Directive 2018/851/EU.

Article 3 of Directive 2008/98/EC, dealing precisely with "definitions," specifies that "waste" is defined as *"any substance or object which the holder discards or intends or is required to discard";* that "holder of waste" is defined as the natural or legal person in possession of the waste or the producer of the waste and that the latter is defined as "the person whose activity produces waste (initial producer of waste) or anyone who carries out pre-processing, mixing or other operations that have changed the nature or composition of such waste."

In Article 6, the same directive, (as amended by 2018/851/EU), specifies what is meant by "end of waste" status: *"Member States shall take appropriate measures to ensure that waste that has undergone a recycling or other recovery operation ceases to be waste if it meets the following conditions: (a) the substance or object is intended to be used for specific purposes; (b) there is a market or demand for that substance or object; (c) the substance or object meets the technical requirements for the specific purposes and complies with existing legislation and standards applicable to products; and (d) the use of the substance or object will not lead to overall adverse environmental or human health impacts."*

The EU legislation therefore, speaks of "end of waste" process defining in the negative (denying the nature of waste), the outcome of the valorization process and not defining in the positive (naming) that outcome. The matter subject to valorization according to the criteria of Article 6 of 2018/851/EU, being no longer waste, returns to being a good[5] again, or a second raw material. The concept of second raw material is not a legal definition found in EU legislation.[6] And according to some,

---

[5] Interesting, again, on the subject of definition, to dwell on the expression (good) used to refer to any means, whether tangible or intangible, that is deemed appropriate for the satisfaction of a need.

[6] In Law 475 of Nov. 9, 1988 (conversion into law, with amendments, of Decree Law No. 397 of Sept. 9, 1988, containing urgent provisions on the disposal of industrial waste), the word "secondary raw materials" appears: Art. 2. - (Secondary raw materials). - a. Secondary raw materials are residues resulting from production processes and which are susceptible, possibly after suitable treatment, to be used as raw materials in other production processes of the same or other nature. In EU documents, an explicit reference to the concept of second raw material is found in Communication COM(2015) 614 final - Closing the loop - An EU action plan for the Circular

the meaning used is not intended to affirm a value but, on the contrary, *"as it has always been used, the concept of second raw material is intended to denote a dis-value rather than a value: it is a term used to "tag" a product and make it second-class because it originates from recycling processes"* (RONCHI). It would be interesting and useful to have its own name to distinguish it from virgin matter, emphasizing the value of its origin. In fact, "*to say that a thing is determined means that we attribute it to a class of commercial objects and also that we give it a name that allows it to be easily determined by attributing it to this class*" (Boltasky, Esquerre 2019 p.126).

8.3 The Explanation of PET and r-PET Prices According to the Respondents

When asked "What do you think determines the price of virgin PET?", most respondents[7] (10) said the price of oil, followed by market dynamics, i.e. the interaction between supply and demand (8). With respect to this aspect, it was pointed out that PET is a material that in recent years has shown some stability (price-wise) compared to other polymers. In fact, in the case of other polyolefins, the market is concentrated on a few producers; in the case of PET, on the other hand, supply is on average very high and fragmented, while consumption has had some increases but no peaks (ALIPLAST). However,

> *during the 2020 lock down, demand for PET was very low, and the price dropped to 600 €/ton. This year (2021) demand has sharply increased and the price has risen to 1,3000 €/ton. In any case, since the prices of different plastics change day by day, week by week, is to make an analysis from the average price or the point price (PLASTIC CONUSLT).*

Only 4 subjects very accurately indicated the influence of monomer prices from which one starts to make PET polymerization. It is no coincidence that these subjects are among those closest to the petrochemical plastics industry.[8]

> *What impacts most on the final pricing of PET is the price of the individual monomers, which are TPA (terephthalic acid) and ethylene glycol (EG), which itself is an antifreeze (PLASTIC CONUSLT).*

And prices of individual monomers fluctuate constantly:

Economy, in paragraph 4 "From waste to resources: boosting the market for secondary raw materials and water reuse".

[7] Not all respondents answered because they did not consider the question to be within their competence.

[8] A consulting firm and its cultural spin-off, the Institute for the Promotion of Plastics for Recycling, whose founding member is UNIONPLAST, as well as COREPLA; and a chemical recycler, whose goal is to make monomers that are competitive with virgin ones.

there are even different supply contracts, yearly, or monthly, to mitigate price fluctuations, with specialized companies tracking week-to-week trends. Everything is tied to the cost and availability of oil. (IPPR).

GR3N, a chemical recycler, tries to sketch a ratio:

Let's say that the final price of virgin PET is 1,000 euros per ton, and that this in turn is given by: cost of conversion from monomer to polymer (about 200 euros); a 5% margin on the total cost of virgin (50 euros). This means that the cost of a basic monomer is around 750 euros or so. GR3N's goal is to produce monomers to be sold at a price similar to that of virgin monomers.

And three parties explicitly counted production factors among the factors that determine the final price of virgin PET, although a marginal influence of the latter is acknowledged, "since the technology for production is now well established" (RONCHI) (Fig. 8.3).

When asked "What do you think determines the price of r-PET (from mechanical recycling)?", most respondents[9] (9) said market dynamics, i.e., the interaction between supply and demand. Five subjects focused on different types of costs: collection (5), sorting (5), production (3).

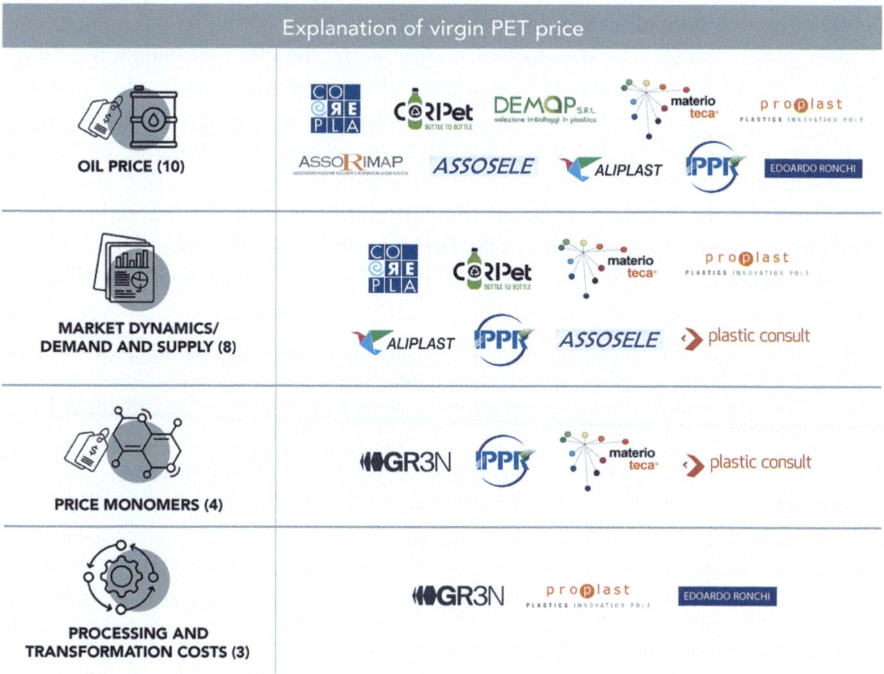

Fig. 8.3 Explanation of virgin PET price according to interviewees

[9] Not all respondents answered because they did not consider the question to be within their area of expertise.

Next, the relationship with the price of virgin PET (7), believed to be effective until recently, and now no longer, and a source, therefore, of destabilization in the system was reported as a motivation.It is interesting to start from this last consideration by listening to the voice of the interviewees.

> *A few years ago I would have said that what determined the price of r-PET was virgin PET (r-PET cost less than virgin); but today I recognize that this is no longer true* (DEMAP).

> *R-PET has always been cheaper than virgin because it is of lower quality; now market trends say it is starting to be more expensive: there is a more "green" focus* (CGR).

> *In previous years r-PET was cheaper than PET and it worked, because the quality of r-PET was good though not equal to virgin, and therefore the economic convenience made up for not having a first choice. In recent years, the appreciation from policymakers and the companies themselves has gone up* (ASSORIMAP).

> *Until two years ago, recycled material, was called cost saving: being of worse quality to virgin had a lower cost* (GARBO).

> *Mechanically recycled PET is always interconnected to virgin PET. Recycled PET is valuable: high quality, more so than other mechanically recycled polymers. Legislation push in the direction of using it, so it makes it a further valuable product* (PROPLAST).

It could be said that until recently virgin PET served as a "meta-price, that is, it supported a price justification for recycled PET (Boltansky & Esquerre, 2019). Then something happened that blew the fit. Regulation, acting as an agencement, created shifts, changed the structure of reality, with "the effect of enhancing things hitherto considered of secondary importance or ignored altogether, and in this way raising their price relative to other objects" (Boltansky & Esquerre, 2019 p. 155).

Through three time steps, it is possible to reconstruct this dynamic of gradual increase in the price of r-PET vis-à-vis PET from mid-2018 to mid-2021. In the July 2018 Varese news article[10] we read about Ferrarelle's new bottle2bottle plant[11]: "The cost of a ton of used bottles costs between 300 and 650 euros, depending on the characteristics of the bottle, compared to 1,250 euros for a ton of new PET." In mid-January 2021,[12]

> *the quotations are 700-800 €/ton for r-PET, from mechanical recycling; 800 €/ton for virgin PET; 1,300 €/ton the purified granule (chemical recycling). There is increasing competition for r-PET, on the one hand, and there is surplus of virgin PET on the other"* (ROBOPLAST).

> *As of April 2021, a bale of post-consumer plastic (with about a 10 percent unsuitable fraction in it) came to cost €800/ton. Taking into account that mechanical recycling has a recovery yield of 70-75%, it's easy to find the final price of r-PET, it's about €1400/ton (*

[10] Presenzano plant in the province of Caserta: the foundation stone was laid in 2015, and it became operational 2019. However, only since 2020 has it been traveling at full capacity, with 18,000–20,000 tons recycled annually.

[11] https://www.varesenews.it/2018/07/produrre-bottiglie-plastica-riciclata-conviene/738676/

[12] https://www.icis.com/explore/commodities/chemicals/polyethylene-terephthalate-recycled/europe/ *Site with service and access fee. ChemOrbis is the other site where chemical materials price lists can be found. In both cases are reports are paid. Roboplast has an annual subscription.*

r-PET flake), broken down as follows: about €1,145 (800 divided by 70%) for the raw material; about €300/ton for the production cost; if I want to make it suitable for food count it takes another €200-300" (GARBO). In the same period, virgin is around 1,300 €/tonn (PLASTIC CONSULT).

FERRARELLE's statement about the lower cost of virgin PET compared to r-PET is interesting:

To date, it is still not possible in Italy to make 100% r-PET because there are strong lobbies for virgin plastic that prefer the use of virgin plastic because it costs less. To date, like FERRARELLE, they still cannot pass on the cost of recycled PET to consumers, because those who do not use recycled plastic have a lower cost, and therefore come out in the market with a more competitive price than ours.

The dynamics of the market as a balance between supply and demand, as we mentioned, was indicated by 9 subjects, of which 6 highlighted the scarcity aspect of the supply of the material collected, sorted and sent for recycling, and 5 highlighted the increase in demand. Regarding scarcity, it is explained in these terms, "r-PET suffers from the scarcity of raw material input from post-consumer streams for which, being few, auctions are held to grab the material" (FERRARELLE).

Making supply scarce is precisely the growth of demand, given by the stimulus of regulation, big brands and consumers themselves:

Big giants such as Coca Cola, Nestlé are willing to pay more for r-PET than virgin, and to pass this extra on to consumers. However, this "green" focus of our times is only on post-consumer: whereas before pre-consumer and post-consumer material cost the same (indeed, the former was a bit more valuable than the latter but the smaller volumes reduced its cost) today the trend of considering the enhancement of post-consumer material as something beneficial to the environment means that it is more sought after than pre-consumer material and therefore comes to cost more. (CGR).

Finally, we come to the issue of cost.

About 320 €/ton is the fee paid to municipalities for the separate collection of the plastics fraction (quini not only for PET, because everything is collected); about 200 €/ton is the cost of labor and equipment for sorting (COREPLA).

Interesting in this regard to investigate on what parameters these costs have been identified, and whether they are considered adequate. Sometimes, in fact, the price at which something was paid is disputed, and "the price dispute is always based on a difference in power between seller and buyer" (Boltansky & Esquerre, 2019 p. 138).

If you hear the municipalities say it is inadequate! There has never been a calculation of fees based on a minimum scientific methodology. They did an evaluation of the plastic collection system on the territory of the Metropolitan City of Turin in 2016[13]: the City of Turin collected very little and therefore the cost per ton was huge; and it also collected badly. So it was necessary to do an additional per-intermediate step to do proper sorting, which involved an additional cost. In this case the fee is not adequate at all; on the contrary in the Chierese context the fee is adequate (CTM).

[13] http://www.cittametropolitana.torino.it/cms/risorse/ambiente/dwd/rifiuti/Osservatorio_rifiuti/Rapporto_rifiuti_2016/2_RD_Appendice_imballaggi_plastica.pdf

Effectively, COREPLA has been accused of not granting adequate consideration to municipalities (ACV, 2013; AGCM, 2016). And as for CORIPET, they do not seem to have imagined a new and different way of setting fees than COREPLA, except in recognizing 6 €/tonn more.

There is also no shortage of criticism from the Selectors.

> *Assosele's battle has always been to bring to light the unequal treatment in the selection contract. The selection contract is one size fits all. There has always been a struggle, but unfortunately, there is little impact: there is a monopoly situation, one has always had to accept the impositions of COREPLA. Over time, it was discovered that there are different types of sorting facilities (better and worse). This meant that with respect to the single sorting contract, it was possible to make addenda, i.e., to start experiments (e.g., sorting tubs, yogurt pots....) with those who operated better. Assosele does not want to be paid for the value of the material (the ownership of what goes in is either of the municipalities or the consortia), we want to be paid for the work done for the selection.* (ASSOSELE).

We come, finally, to production costs, and thus to the recycling process. Not all polymers are considered equally recyclable. In fact, every polymer, individually taken, is potentially recyclable. Therein lies the difference: between potentially recyclable and actually recycled. For a material to be said to be recyclable, it is necessary that it be properly collected and sorted and that there be an outlet market, with prices such that it can be purchased. This means that the economic issue is related, upstream, to a question of volumes of material collected (a recycling plant to be sustainable must process at least 20 thousand tons of waste per year) and, downstream, to a question of demand-there must be demand for that recycled good.

As recycling techniques, including chemical recycling, and sorting techniques are developed, it will be possible to process more and more manufactured goods, including those composed of different polymers, as long as these technologies do not cost more than the economic benefit they produce. With regard to PET in particular, a major incidence in production costs relates to the end application to which the recycled PET is directed, and thus whether flake or granule r-PET is needed. *"To make a new bottle I need to start again from granule. To make a stack or tray I can stop at flake. This results in different costs: making granule or flake. Going back to the granule costs more"* (PLASTIC CONSULT).

Again, as with harvesting and sorting, a comprehensive costing exercise that takes into account every aspect and feature of the processing process (capital costs, material costs, labor costs, energy costs, storage, packaging and transportation costs) is necessary to provide a complete and accurate assessment of economic feasibility (Forrest, 2016) (Fig. 8.4).

Fig. 8.4 Explanation of r-PET price according to interviewees

8.4 The Justification for PET and r-PET Prices According to Respondents

Finally, coming to the last point of the triad proposed by Boltansky and Esquerre (2019), let us analyze if and what value appeals were fielded by the interviewees as a justification device for the previously explained prices. Specifically, they were asked which of the two prices (virgin PET and r-PET) they think is right to be higher and why.

Ten subjects argue that the price of virgin PET should be higher, four the price of r-PET, three say the question does not make sense, and one subject believes they should be equal. Several points deserve attention in light of the answers given. First is the belief that such a question is meaningless. For two out of three subjects the nonsense comes from the fact that the market decides the price (neoclassical and marginalist school of economic thought). There is not the slightest consideration given to the fact that we can question the need for a fair price with respect to things. *"Price is not something abstract; it depends on the dynamics of supply and demand. There is no reason why one or the other has to be right"* (COREPLA). As if reasoning around the question of a fair price is something abstract. The other subject, on the other hand, points out that it makes no sense to determine a price a priori, because this depends on several conditions, including production costs and the product's performance capabilities.

A second striking aspect is that one subject may consider several justifications, among them alternatives, to be true at the same time. Such as COREPLA, which declares the nonsense of the question, and then adds, *"If we make a cost argument: it costs more r-RET, and therefore has a higher price; if we make a performance argument it is worth more virgin (and therefore costs more).* Not contemplating the possibility of reasoning around the right price, in this case, the concept of "environmental cost" was not considered until after it was explicitly fielded.

> *To date, the environmental cost is not part of the final price definition; certainly if it were counted, then it would cost more for virgin PET because of the use of virgin resources and the cost of energy for extraction, refining, transportation. In this regard, a WWF Report*[14] *recently came out that tries to quantify the real social and environmental cost of virgin plastic production. However, the result that comes is sheer madness: $3.7 trillion; it is unclear on what basis they made those calculations.* (COREPLA).

Six parties in addition to COREPLA argue that the price of virgin PET should be higher than r-PET by spontaneously justifying the choice with the concept of environmental cost.

> *The benefits of using recycled, waste, and non-virgin material must translate into benefits for the future (less environmental damage, less pollution preservation of biodiversity...) (GARBO).*

> *Today, for our consumption model we pay the price of the product and what is around it (the whole chain: from conception to distribution) but the environmental cost is not paid* (CMT).

Paradoxically, the same concept can be used to motivate a higher price of r-PET:

> *In an ideal world, where everything is 30 percent recycled, the recycled bottle costs more than a 100 percent virgin plastic bottle because you have to be willing to pay more to preserve something that needs to be maintained into the future. It is the consumer who has to be willing to pay more. We are in a world where competition does not exist and free-market distortion prevails.* (FERRARELLE).

Interestingly, the justification regarding willingness to pay can be adopted either by those who argue that virgin PET should be higher (willingness to pay for higher product performance); or by those who argue that r-PET should be priced higher (willingness to pay for the built-in environmental benefit). In this second case, it is striking how the previous discussion regarding environmental cost, translates into

[14] https://www.wwf.it/pandanews/ambiente/inquinamento/il-caro-prezzo-della-produzione-di-plastica/

terms of the environmental benefit that goes into the product's plus and is paid for by those who want/can afford it.

And we come to the last point worthy of observation: just as the same value convention (willingness to pay) can justify two opposing scenarios, so can the same scenario within it be justified by opposing value conventions. This is the case related to the discourse setting of price justification in terms of reward or penalty. Indeed, it is possible to justify a higher price of virgin PET by emphasizing a penalty character (internalizing the environmental cost; making it less competitive against r-PET); as well as, conversely by emphasizing a reward character (higher value, because of higher quality and performance). Similarly, it is possible to justify a higher price of recycled PET by emphasizing a penalty character (the consumer bears the cost of conservation); as well as, conversely by emphasizing a reward character (giving oxygen, value and prestige to the recycled PET supply chain).

In light of these considerations, it is possible to make a matrix to visualize the co-presence of even antagonistic value conventions within price justification discourses (Figs. 8.5 and 8.6).

All those who have recognized the need to incorporate environmental costs within virgin PET production, likewise critically believe that the actual environmental benefit of recycling should be constantly verified.

Giorgio Nebbia (2002, p. 143) explains this concept well: the environmental cost of each commodity or service, can be defined as "*the quality of energy, or the mass of water, or minerals, or vegetables, or the mass and type of waste that accompany the production or use of a unit of weight of commodity or a unit of service*." Indeed,

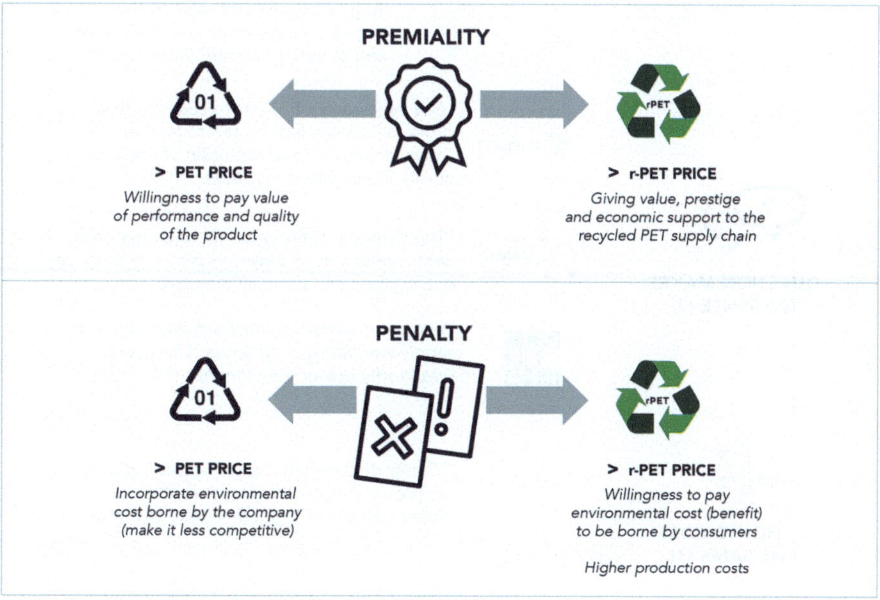

Fig. 8.5 Premilitary/penalty matrix

Fig. 8.6 PET and r-PET price justifications

under conditions of scarcity, it is all the more useful or appreciable (and therefore has greater value) to have a process that yields the same commodity and service, with less consumption of raw materials, or with less energy consumption, or with less environmental pollution. However, of energy and matter, one must also evaluate the "commodity" quality, understood as the aptitude to provide, after each transformation, still useful energy, net of losses due to increased entropy.

The invitation, addressed to the interviewees, to think about the issue of a possible fair price for r-PET and virgin PET, in conclusion, wishes to lead to the recognition that price is a "constructed fact," which does not mean denying objectivity, but recognizing the proper role of each actant involved (human being, natural element, regulatory and technological device whatever) (Latour 1986).

8.5 Conclusions

In this paper we investigated the recycled PET supply chain in connection to the "right" price of recycled PET and that of virgin PET. The crux behind this question is to understand whether supply chain actors recognize the issue of the environmental cost inherent in the production of any good or service, and thus whether they recognize the possibility of anchoring the price to a *value* that is not determined solely by the meeting of supply and demand. Giving a price is one of the possible forms of valuing and assessing a good: the price, like the mask, reveals and conceals at the same time. Boltansky and Esquerre (2019, p. 132) identify three foundational elements the form of valorization that goes through assigning a price: initiating a valuation process, having a shared measurement system, contingent character of the event.[15]

First, assigning a price, as already mentioned, means carrying out a valuation process; however, valuation may not always be reduced to a matter of monetary price alone (Stark, 2011). Is it possible, then, to identify a price that is not expressed in monetary terms? And how can such a price be established? Traditional economic policies assume that affecting prices and assigning property rights is an effective lever for changing and conditioning the behavior of people and/or firms. What matters is defining the "right price." What does it mean to define them, and keep them "fair"? The economic system we are immersed in leads to overestimating the role of prices and underestimating the role of value judgments and how some things are put at risk once a monetary price is assigned to them (Sandel, 2013). In fact, setting a price is: "like rubbing a match: it catches the eye, but the sparks can be dangerous" (Raworth, 2017, p. 132). When we decide that some goods can be bought or sold, it

[15] As much as the authors have in mind a very particular contention (physical products and objects, related to the world of antiques, used goods, design objects), we think it is interesting to adopt the three elements they indicate by broadening the discourse to a different variety of goods, not just commodities, and where the assignment of price, does not necessarily have to take on a monetary value, as much as the recognition of a value that needs to be taken into account.

means that we ascribe to them *the status of a commodity*. But not all goods, in this way, are valued correctly. Indeed, there are several goods that by ontological status cannot be possessed and by the mere fact of being used, create ties, presuppose ties. A gift, for example, may not even take the form of a material good, but rather an immaterial good, configuring itself as relational goods (Gui, 2002). In this domain, natural goods are also rightly found. For all these goods, it is possible to recognize a "constitutive incommensurability" (Raz, 1986), that is, an incommensurable value, precisely because it is not possible to possess them, let alone eradicate them from the network of relationships in which they are immersed (Fig. 8.7). Quite rightly, for these goods one can speak of "social use value" (Kapp, 1963), and environmental, we would add, a definition of value that suggests and is based on issues of preservation of environmental balances and satisfaction of the basic needs of human being.

In this line, a special consideration deserves natural resources, compared to other goods characterized by "constitutive incommensurability," which have a peculiarity: being input and condition for the development of productive economic activities. Certainly having for too long considered natural resources to have zero economic value, and thus zero cost, has led to problems of overuse of them and overproduction of waste. And in the facts, this translates into a valuation error rather than a lack of valuation. In fact, *"although ecosystem assessment is certainly difficult and full of uncertainties, a choice we do not have is whether to do it or not. Rather, the decisions we make as a society about ecosystems always involve valuations (though not necessarily expressed in monetary terms). We can choose to make these evaluations explicit or not; we can make them with an explicit recognition of the enormous uncertainties involved or not; but as long as we are forced to make choices, we are going through the evaluation process"* (Costanza et al., 1997, p. 255).

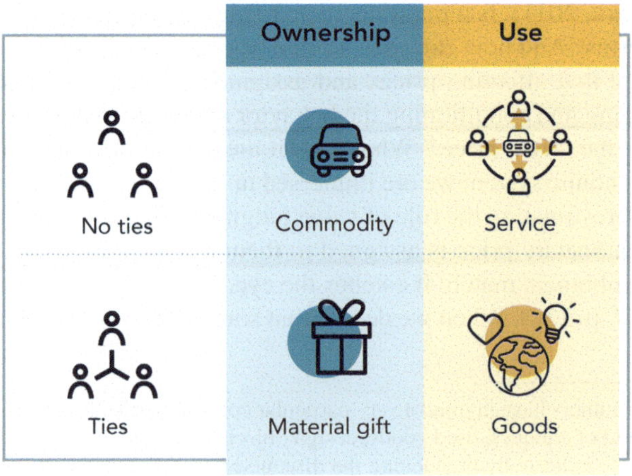

Fig. 8.7 Goods circulation matrix (own elaboration)

This is why, while acknowledging all methodological limitations, it can be instructive to estimate the "incremental" or "marginal" value of ecosystem services. Indeed, *"if ecosystem services were actually paid for, in terms of value to the global economy, the global price system would be very different from what it is today. The price of goods that directly or indirectly use ecosystem services would be much higher. The factor payment structure, including wages interest rates and profits would change dramatically. World GNP would be very different in both magnitude and composition if it properly incorporated the value of eco-systemic services (Ibid., p. 259).*

Which means working to find modes of valuation such that the price of goods does not merely figure as the meeting point of supply and demand, but can also take into account the value (defined from a noneconomic valuation process) inherent in resources. In this regard, Nebbia (2002, p. 12) looks forward to the "*day when goods (and services) will be sold not according to a price in euros or dollars or yen, but according to the mass of raw materials used, or the amount of energy required.*" In fact, "*in the light of the problems of scarcity of certain factors provided by nature (not only oil or forests, but also the receptive capacity of rivers and air) goods and services that, at the match of economic utility, require* fewer *raw materials, less energy,* last longer, *generate* less *waste* , *involve* less *pollution and* less wear and tear on *natural resources are* worth more" (Nebbia, 2002, p.55). The claim is, therefore, to develop plural ways of assessing the environment that can recognize and value, without raping, a "constitutive incommensurability" of natural resources, whose career, when they enter production cycles, leads them to become commodities.

Several researches are now trying to analyze through LCA analysis the different environmental impacts generated by the production of virgin PET, r-PET, and even bioPET (Ramesh, Vinodh, 2020). With regard to the research questions, then, (justification of virgin PET and r-PET prices), the paper showed how to work out a new plural way of valuing plastics production and the environment by tying PET and r-PET prices to LCA analyses and defining: "*in what terms economic valuation is linked to other non-economic forms of valuation* (Centemeri, 2015, p. 303). In this way, the values expressed by LCA analyses can be elected as a meta-price (Boltanski, Esquerre, 2019) of the current prices of virgin PET and r-PET, acting, thus, as a device for critiquing those prices. Hence serving as references for determining a minimum price threshold below which it is not possible to go, thus banishing environmental (and social) dumping practices.

We need, then, other forms of evaluation and social and political determinations of "social use value" (Kapp, 1970), involving different forms of knowledge about the environment and the interactions humans have with it. This aspect leads to the second element characterizing a value assignment process, namely, that it is necessary to have a shared measurement system. There are two main characteristics of a shared measurement system: being able to deal with moral uncertainty (conflict between different value orders and commitment regimes), and simultaneously with epistemic uncertainty (inability to identify the specific causes of the incidences of environmental and social harms because they are immersed in complex systems based on social and ecological interdependencies) (Centemeri, 2009). The plurality of ways of defining what constitutes value, in fact, goes hand in hand with the different desirable collective goals to be pursued: allocative efficiency is only one of them. Therefore, "*analyzing the moral limits of markets means reasoning collectively and publicly about how to value the social, relational, and natural goods to which we put a price*" (Sandel, 2013, p. 18). To continue to deny ourselves, as we have done so far, such public debate condemns us to being a market society, instead of being a multidimensional society that has, among others, an effective tool in organizing productive activity, such as *having* a market economy (Sandel, 2013).

This depends on the concrete organization of the decision-making processes that govern economic supply chains, the interests, knowledge, property rights, and powers involved. The organization of decision-making processes defines the ability to hold together past solutions that provide certainties with new and uncertain ones into a single innovation strategy (Sabel & Victor, 2022). The development of value-creation mechanisms that do not reduce the forms of value to market price thus requires an experimentalist and heterarchical governance, characterized by multiple, overlapping evaluation principles rather than a single, fixed hierarchy. Unlike traditional hierarchical systems, where authority and decision-making follow a clear chain of command, heterarchies operate through the simultaneous presence of multiple logics or criteria for evaluation (Stark, 2009).

This would allow to address not only the pluralism of modes of assessment, but also the "radical uncertainty" and ignorance that are characteristics, at the epistemic level, of environmental issues (Centemeri, 2015).

As a policy suggestions, our research points to the possibility to establish *hybrid forums* (Callon, 1998; Callon & Law, 1995) in which the different actors of the PET value chain (from packaging producers to citizens, from waste handlers to recyclers, from legislators to consortia, to designers) *introduce* both epistemic and moral uncertainty in the decision-making process. This would define new logics of worth for the environmental cost of resource and energy use to be adequately reflected in the final price of virgin PET and de r-PET. It is key to point out that, as the experimentalist governance framework clearly shows, both technical and non-technical dimensions are key. In this line, we showed the role of the LCA techical analyses, which is also subject to a non-technical evaluation with respect to the parameters used. This points to the key role to define ways of establishing and evaluating the various environmental certifications[16]; and, again, define collegial processes of formulating specific laws, such as the plastic tax, and in general the possibility of taxing the use of virgin material (Bruvoll, 1998) and the transposition of EU directives.

Finally, the third element of a value assignment process concerns the contingent nature of value. Valuation is always spatially localized and temporally marked (Hutter & Stark, 2015). Once a shared valuation process is defined that links

[16] The Ecolabel criteria, established at the European level, already provide for the participation of stakeholders including European consumer and environmental associations to discuss various issues including also consumer health and safety and social and ethical aspects of production processes. (https://www.isprambiente.gov.it/it/attivita/certificazioni/ecolabel-ue)

monetary and nonmonetary dimensions, the definition of a final price of a good, can also be conditioned by the dynamics of locally and temporally situated supply and demand matching, dynamics that, therefore, is one of the components of the determination of a final price, but, precisely, not the only one determinant.

References

AGCM (2016) *Indagine conoscitiva sui rifiuti solidi urbani*. https://www.agcm.it/dotcmsDOC/allegati-news/IC49_testoindagine.pdf

Associazione Comuni Virtuoisi (2013) *Per un nuovo accordo quadro ANCI-CONAI*. https://comunivirtuosi.org/campagne/anci-conai/il-dossier/

Boltansky L., Esquerre A. (2019, ed. orig. 2017). *Arricchimento. Una critica della merce*. Bologna, Il Mulino.

Bruvoll, A. (1998). Taxing virgin materials: An approach to waste problems. *Resources, Conservation and Recycling, 22*(1998), 15–29.

Calıskan, K., & Callon, M. (2010). Economization, part 2: A research programme for the study of markets. *Economy and Society, 39*(1), 1–32.

Callon, M. (1998). An essay on framing and overflowing: Economic externalities revisited by sociology. In Id (Ed.), *The laws of the markets* (Vol. 46, pp. 244–269). Blackwell Publishers.

Callon, M., & Law, J. (1995). Agency and the hybrid collectif. *The South Atlantic Quarterly, 94*(2), 481508.

Centemeri, L. (2009). *Environmental Damage as Negative Externality: Uncertainty, Moral Complexity and the Limits of the Market* (pp. 21–40). e-cadernos CES 05 | 2009 As fundações institucionais da economia.

Centemeri, L. (2015). Reframing problems of incommensurability in environmental conflicts through pragmatic sociology: From value pluralism to the plurality of modes of engagement with the environment. *Environmental Values, 24*(3), 299–320.

Costanza, R., d'Arge, R., de Groot, R., Farber, S., Grasso, M., Hannon, B., Limburg, K., Naeem, S., O'Neill, R. V., Paruelo, J., Raskin, R. G., & van den Belt, M. (1997). The value of the world's ecosystem services and natural capital. *Nature, 387*(15), 253–260.

Hutter, M., & Stark, D. (2015). Pragmatist perspectives on valuation: An introduction. In A. Antal, M. Hutter, & D. Stark (Eds.), *Moments of valuation: Exploring sites of dissonance* (pp. 1–14). Oxford University Press.

Kapp, K. W. (1970). Environmental disruption and social costs: A challenge to economics. *Kyklos, 23*(4), 833–848.

Lacy P., Rubqvist J., Lamonica B. (2016, ed. orig, 2015). *Circular economy. Dallo spreco al valore*. Milano, Egea.

Nebbia, G. (2002). *Le merci e i valori. Per una critica ecologica al capitalismo*. Milano, Jaca Book.

Ragaert, K., Delva, L., & Van Geem, K. (2017). Mechanical and chemical recycling of solid plastic waste. *Waste Management, 69*, 24–58.

Raworth, K. (2017). *L'economia della ciambella*. Sette mosse per pensare come a un economista del XXI secolo. Edizioni Ambiente.

Raz, J. (1986). *The morality of freedom*. Calrendon Press.

Sabel, C., & Victor, D. (2022). *Fixing the climate: Strategies for an uncertain world*. Princeton University Press.

Sandel M.J.. (2013, ed. orig. 2012). *Quello che i soldi non possono comprare. I limiti morali del mercato*. Milano, Feltrinelli.

Star, S. L. (1995). The politics of formal representations: Wizards, gurus and organizational complexity. In S. L. Star (Ed.), *Ecologies of knowledge: Work and politics in science and technology* (pp. 88–118). SUNY Press.

Stark, D. (2009). *The sense of dissonance: Accounts of worth in economic life*. Princeton University Press.
Stark, D. (2011). What's valuable? In J. B. E. P. Asper (Ed.), *The worth of goods. Valuation and pricing in the economy* (pp. 319–338). Oxford University Press.

Lambiase Nadia University of Turin Ph.D. in Innovation for the circular economy, currently is assistant professor at the University of Turin, where carries out research on drivers and barriers to the ecological transition of companies. Socially-minded entrepreneur, CEO of Mercato Circolare, a benefit company operating in Italy.

Filippo Barbera is Professor of economic sociology at the CPS Department of the University of Turin and Fellow at the Collegio Carlo Alberto (Torino). He is member of the Foundational Economy Collective and of the Forum Inequalities and Diversities. His recent publications include: Commons. Citizenship and Power: Reclaiming the Margins, Policy Press, 2025 (eds. With E. Bell). Currents research projects focus on the role of democratic populism in marginal areas, the regeneration of the public sphere and the analysis of foundational economy experiments in the provision of citizenship goods and services.

Chapter 9
Geopolitics of Circular Economy

Massimo Franchi

Abstract The first Industrial Revolution gave rise to an economic system, born with the use of the steam engine and centered on the factory, based on the logic of market saturation and consumption linked to the satisfaction of basic needs. Over time, the West, thanks to electricity and the internal combustion engine, saw the rise of a process marked by the cycle of mass production and its disposal, which, in the modern era with the advent of information technology and the saturation of markets, shifted the focus to the economy of services linked to the product with the goal of gaining customer loyalty and the satisfaction of new dormant needs. The effects of this growth model have had significant environmental and social impacts, such as mass consumerism and widespread pollution of air, land, and water. In the East and in models other than western capitalism, things have not been any better. Where planned economy has prevailed, environmental and social damages have been far greater, due to a lack of sensitivity to the control of public affairs, while Developing countries have seen development with lack of a culture of civic responsibility that has led to a disposal of products that has gone from initial reuse, due to scarcity, to reckless disposal failing to foresee the impacts on the environment and people's health, as well demonstrated by the amount of plastic that has accumulated in the Oceans. Nowadays, there is widespread awareness not only of the interconnection between the economy and the environment, but also that an economic model, based on continuous growth built on the consumption of scarce resources, has increased pressure on the environment, generating negative externalities that threaten ecosystems and deplete natural resources. Globally, the inefficiencies of different economic systems not only show that many resources are wasted—for example, in developed Europe only a fraction of waste is recycled and in the rest of the world things are much worse—they also condemn future generations to pay a very high debt not just in economic terms, but also in relation to quality of life and opportunity. A circular economy, on the other hand, could provide not only a useful approach

M. Franchi (✉)
University of Modena and Reggio Emilia, Modena and Reggio Emilia, Italy
e-mail: mfranchi@unimore.it

S. Patnaik et al. (eds.), *Circular Economy and Sustainable Value Creation through Eco-Innovation*, Circular Economy and Sustainability,
https://doi.org/10.1007/978-3-031-88516-7_9

to eliminating waste by valuing all the resources already available through maximum value extraction, but also a criterion capable of generating new value as, for example, in the case of second-hand clothes that are collected from waste, sorted, assessed, arranged and repurposed for "vintage" sale, producing new value and transforming the very concept of the garment. Therefore, circular economy is not only about decreasing the consumption of raw materials, but also about designing sustainable products with a longer lifecycle, which includes their reuse through remanufacturing and finally recycling. In a nutshell, a change in mentality is suggested, through an ethical choice, both in evaluating the durability of products and in conceiving the purchase of something that is not new. Such a transformation clashes with the concept of market share, continuous growth and the company's short-term focus. It is clear that in order to overcome these cognitive hurdles, it is crucial that companies internalise sustainable growth into their strategy and, even before that, as part of their purpose. As proven by a number of successful cases of companies operating with the circular economy philosophy, it is now a proven fact that it is able to generate growth and income, while at the same time reducing raw material waste. The process of conversion to circular economy, also carried out in Europe by automotive manufacturers in the premium segment, not only mitigates supply risks, but also the trade deficit of processing countries such as Italy or Germany, generating jobs thanks to reconditioning, which in some sectors involves the employment of labor with relevant technical skills. This is an issue that is also becoming geopolitical, given the different positions of different governments in pursuing sustainability and waste reduction policies. As in the case of CSR, there are governments that are investing heavily in the reuse and remanufacturing of products and others that are struggling with waste management, which has become a real ecological bomb. The very choice to dispose of waste in other countries, which are able to propose appropriate and functioning technologies and solutions, changes the relationship between governments and proposes projections of influence that can be decisive in the near future. The transition to electric in automotive itself is generating significant issues in relation to battery disposal and the sustainability of the business, given the use of raw materials for construction that heavily affects the planet. This paper will analyze the geopolitical and geoeconomic impacts of circular economy and how it can become a force in the foreign policy of nation states.

Keywords Economy · Market · Circular economy · Geopolitics · Strategy · CSR · Value · Foreign affairs · Enterprise · Environment · Citizen · Law · Development · Geoeconomics · Economic system · Industrial revolution · Pollution · Raw material · Poverty · Economics' human rights · New generations · Foreign policy · Spheres of influence · Innovation · Transformation · Automotive

9.1 Introduction

The first Industrial Revolution gave rise to an economic system, born with the use of the steam engine and centered on the factory (Weber, 2010), based on the logic of market saturation and consumption linked to the satisfaction of basic needs (Sombart, 1978). Over time, the West, thanks to electricity and the internal combustion engine, saw the emergence of a process characterized by the cycle of mass production and its disposal, which, in the modern era with the advent of information technology and the saturation of markets, shifted the focus to the economy of services linked to the product with the goal of customer loyalty and the satisfaction of new dormant needs. Re-using, recycling, repairing, saving: the use of these words is not just an exercise in recalling terms familiar from the past, when the re-use of materials and objects was an everyday practice. It contrasts with the linear logic summarised by the triad 'take-produce-dispose' typical of the traditional economy. Such logic involves the satisfaction of new needs as the main objective of many companies, with increasingly innovative products whose useful life has shortened over time. In the past, an artefact was made to last as long as possible according to the logic of natural ecology, which has led to clear examples that have come down to us, such as Roman bridges and aqueducts. In modern times, consumer fashion has imposed increasingly tight deadlines and set a disposable mentality contrary to saving and ecology. A mentality which, paradoxically, has also affected the construction of infrastructures whose duration and quality have decreased over time despite using new techniques and raw materials. Fortunately, it is now understood that reusing, recycling and recovering saves money and other resources, such as labour, physical energy and time. From the first industrial revolution onwards, especially after the saturation of elementary needs, these concepts were set aside by the search for constant and infinite growth shaken by periodic economic crises. An example of recovery is well represented by the churches often built in previous times, in which it is possible to understand the different eras and the recycling deriving from previous structures by studying the layers of the different materials used (Bruni, 2015). The consequences of this model of development have seen significant environmental and social impacts, such as mass consumerism and widespread pollution of air, land, and water (Chomsky, 2017). In the East and in models other than Western capitalism, things have not been better. Where the planned economy has prevailed, the environmental and social damage has been far greater, due to a lack of sensitivity to control over public affairs, while in Developing countries there has been a growth without a culture of civic responsibility that has led to a disposal of products that has gone from initial reuse, linked to scarcity, to reckless disposal without foreseeing the impacts on the environment and people's health, as is well demonstrated by the amount of plastic that has accumulated in the Oceans. Nowadays, there is widespread awareness not only of the interconnection between the economy and the environment, but also that an economic model, based on continuous growth built on the consumption of scarce resources, has increased pressure on the environment, generating negative externalities that threaten ecosystems and deplete natural resources (Zingales, 2012). Globally, the inefficiencies of different economic systems not only witness the waste of many resources-for example, in developed Europe only a fraction of waste is recycled and in the rest of the world things are much worse-but also condemn future generations to pay a very high debt in economic terms, but also in relation to quality of life and opportunities (Gaudium et spes, 1965). A circular economy, on the other hand, could provide not only a useful approach to eliminating waste by valuing all the resources already available through maximum value extraction, but also a criterion capable of generating new value as, for example, in the case of second-hand clothes that are collected from waste, sorted, assessed, arranged and repurposed for "vintage" sale, producing new value and transforming the very concept of the garment. Therefore, circular economy is not only about decreasing the consumption of raw materials, but also about designing sustainable products with a longer lifecycle, which includes their reuse through remanufacturing and finally recycling. In a nutshell, a change in mentality is suggested, through an ethical choice, both in evaluating the durability of products and in conceiving the purchase of something that is not new. This is a transformation that clashes with the concept of market share, of continuous growth, and of the company's focus on the short term (Franchi, 2022a). It becomes clear that in order to overcome these cognitive obstacles it is crucial for companies to internalize sustainable growth into their strategy (Coda et al., 2001) and even before that as part of their purpose. Attention to future generations is crucial in the circular economy. It focuses on the problem of using scarce natural resources, the continuous extraction of which is becoming incompatible with long-term sustainability. We are talking about limited resources from which future generations will not benefit. From this perspective, the circular economy is a new model of resilient growth that introduces the theme of safeguarding and the responsibility of leaving new generations a natural legacy that allows them to pass it on to subsequent ones. The custody of creation and the care of the common home (Francesco, 2015) should be considered essential objectives of the human race regardless of origin and religion. For Pope Francis: *"We have not yet managed to adopt a circular model of production capable of preserving resources for present and future generations, while limiting as much as possible the use of non-renewable resources, moderating their consumption, maximizing their efficient use, reusing and recycling them. A serious consideration of this issue would be one way of counteracting the throwaway culture which affects the entire planet, but it must be said that only limited progress has been made in this regard". (Laudato Si' 2015)."* These reflections lead to thinking on the ethical issues underlying the circular economy, particularly the principle of the common good. Simply put, it represents a widespread value whose definition can be summarised as follows: *"Every day human interdependence grows more tightly drawn and spreads by degrees over the whole world. As a result the common good, that is, the sum of those conditions of social life which allow social groups and their individual members relatively thorough and ready access to their own fulfillment, today takes on an increasingly universal complexion and consequently involves rights and duties with respect to the whole human race".* (Gaudium et spes 1965). Actually, however, the prevalence of the interest of a few, linked to logics that combine profit and technology, reduces the common good to a mere appeal and distances it from the fact that the needs and aspirations of the human family are at stake: *"Every social group must take account of the needs and legitimate aspirations of other groups, and even of the general welfare of the entire human family"*. (Gaudium et spes 1965). The concept of circular economy has changed over time. It was related to people's needs and the scarcity of available products in the past. This state could still be attributable to some areas of the planet, even if Western countries with mature and saturated economies need incentives and motivation to transition from a linear to a circular economy (Stahel, 2019). This transition must consider the new generations, who will have to actively participate in rethinking their future. The future should not be determined only by the elderly, whose contribution remains important to avoid mistakes, bring experiences and maintain a high motivation level. The elderly or adults who often propose a work model are still linked to Fordism (Rainieri, 2019), which is not appreciated today by the younger generations, much more interested in creative activities, at least in the West, and less willing to sacrifice relational time (Franchi & Rainieri, 2019).

9.2 Theoretical Background, Literature Review

As demonstrated by a number of successful cases of companies operating with the circular economy philosophy, it is now a proven fact that it is able to generate growth and income, while at the same time reducing raw material waste. The process of conversion to circular economy, also carried out in Europe by automotive manufacturers in the premium segment, not only mitigates supply risks, but also the trade deficit of processing countries such as Italy or Germany, generating jobs thanks to reconditioning, which in some sectors involves the employment of labor with relevant technical skills.

The literature on the subject believes it is imperative to differentiate the political aspect (Molteni & Alessandra 2008), with the push for ecological transition, which in the West has been supported by international organisations such as the EU, from the entrepreneurial aspect, related to the need for companies to continuously evolve their business model. Furthermore, an ethical aspect is fundamental to applying an authentic circular economy but is often used only as a marketing tool.

Each of these three areas can narrate the topic of the circular economy, highlighting only its positive aspects and offering a unique, shared and favourable vision of the world.

Actually, however, many problems and even dangerous drifts could arise. For example, from a political and regulatory point of view, the requests of a block of countries regarding the ecological transition could trigger international competition between states or impose technologies, such as the electric car, whose positive effect in terms of environmental and social impacts is not certain, but they would be necessary to maintain or bridge the technological lead between states.

In the European Union (Malgorzata Jakim et al. 2024), the concept of Open Strategic Autonomy (OSA) highlights the need to act autonomously without depending too much on other states and create strategic global value chains called Global Value Chains (GVCs). The CRM Act, which came into force in May 2024, establishes the benchmarks for promoting the "reshoring" and diversification of Strategic Value Chains (SVCs) by 2030. The threshold of the limit of 65% of imports of raw material from a single country sets a certain boundary not to be exceeded and from which to stay away but also presents a pragmatic clash on the scarcity and location of the same (Regulation (EU) (2024). 2024/1252). In short, it is about combining protectionism and European security with the European Green Deal in the awareness of vast areas of strategic inconsistency. However, if reshoring is considered a solution for the independence of the European Union and to reduce global supply chains and the risks deriving from them, there is no doubt that offshoring has represented a great opportunity for developing countries, China first and foremost, improving the living conditions of hundreds of millions of people. Updating the 2020 new industrial strategy: building a stronger single market for Europe's recovery.

Some scholars have focused on titanium (Jakimów et al., 2024), which is considered a strategic raw material, therefore CRM, for future aerospace innovations and national security, given its past origin in Russia. The reliance on the American industry, which has invested heavily in this raw material, with the movement of the supply chain through "friendshoring", does not protect the European aeronautical industry from risks. For these scholars, "friend-shoring" would be better achieved by promoting a circular economic strategy based on what has been done in the USA through the recycling and reuse of scrap. This is about applying an active method to create new processing plants within the EU. A method that could be used by every country in the world and which would allow savings in resources and safeguard employment.

Then there is the problem of companies that could decide to ride the circular economy only for a marketing strategy aimed at improving their reputation, remedying past bad practices or simply aligning themselves with what is being carried out by competitors (Wookey et al. 2018). Unfortunately, it has happened that this practice has been applied by well-known multinationals, which have ridden the circular economy as a trend of the moment without introducing real changes in the business model (Franchi, 2022b).

The ethical part, which is of great importance for supporting the transformation towards the circular economy, could instead highlight ideological limits and close in on itself, hindering any form of innovation, such as new-generation nuclear power, and imposing a new form of single globalist thought that divides the world between those who are in favour and those who are against it.

According to some researchers (Preston et al., 2019), Circular Economy is becoming a resilient growth model in developed and developing economies. The reason for this spread is to be found in the alternative strategy to traditional industry driven development. Even in developing countries, it can create new jobs compared to traditional manufacturing, whose limitations are self-evident. In this context, one can think of the workforce needed to clean up plastic from rivers and oceans or the labour needed to process waste.

Certainly, even if developing countries require policies that are different from those of developed ones, due to the different degrees of ongoing environmental risk and sensitivity to the issue, it is clear how important it is to invest in partnerships at different levels and the implementation of educational processes. Each state should implement a national circular economy plan across economic sectors based on identified priorities, also through business incentives (Stiglitz, 1997) and thanks to specific financing mechanisms based on real local objectives and not on non-repayable funds.

This plan should be followed by corporate measurement systems that support economic operators in their daily lives and accelerate their "ambition to action" to move from a linear to a circular model (MacArthur, 2024).

The risk of old-fashioned environmentalism, based on the triad "reduce, reuse and recycle", still communicates the linear and unidirectional economic model to be overcome through new concepts, such as "waste equals food" through which the product, specifically redesigned, becomes nourishment and not rejection (Braungart & McDonough, 2002). According to McDonough, it overcomes the long-standing problem of waste management, which plagues modern societies and impacts on the health of the population, through the generation of regenerative effects in which waste (Fassio & Tecco, 2018), overcome by the closed 'growth-decay-rebirth' process, is transformed into food. This approach changes the way waste is treated in today's societies, at least in the West. Over time, waste has been transformed into energy or recycled materials, minimising the need for landfill. These integrated environmental centres often include new planting, creating a green area for environmental mitigation.

The role of these so-called parks, summarised as 'carbon sinks', is to reduce greenhouse gases and particulates from the traffic generated by vehicles delivering to the hub, but also from the waste-to-energy plants, at the head of which is a preselector, that dispose of residual urban waste from separate collection and special waste. Waste and its management have given rise, in developed countries, to a proper industry with sophisticated logistics and integrated services offered to citizens, such as water management, district heating, city cleaning, etc.

An industry that generates economy that can be converted into turnover (Santesso & Sòstero 2004), EBITDA,[1] employment and shareholder value (Friedman, 1987). An economic system now central to advanced economies and based on a dense network of companies controlled by municipalities or listed on the stock exchange, whose boards of directors whose appointments are highly coveted and sponsored by politics.

Changing or improving this system could be complex in the West and perhaps even more difficult in developing countries where, in the past, the solution has been to open landfills or dumping all kinds of residual waste into the sea without any filter. Over time, these practices were followed by the imitation or perhaps the

[1] *Earnings Before Interest, Taxes, Depreciation and Amortisation.*

acceptance by imposition of the Western model, which did not take into account cultural and environmental differences.

Science has proposed a partial vision of the phenomena by focusing on a specific problem but leaving out the repercussions and impacts of the proposed solutions (Vandana, 2023). In the agricultural field, this practice has been evident with multinationals' push to adopt new seeds protected by patents to replace the native ones, which have adapted for centuries to the biodiversity of the territories in which they were developed and improved. Choices which, despite the aim of improving land yields, have reduced the planet's plant varieties, put thousands of farmers in crisis and increased the use of water resources or pesticides.

The search for a balance between innovation and territorial protection will also become fundamental in the near future to prevent the logic of industrial agriculture, protected by patents, from replacing the conservation of rural areas, which instead base their protection on good practices tested over the centuries and capable of allowing the survival of the population in the area.

9.3 The False Circular Economy

The corporate decision to enter the 'green economy', as a strategic choice (Drucker, 2000) to stand against competitors (Kim & Mauborgne 2005), may risk being ahead of its time and tell a very different story from the company's actual situation. This strategic move, which may be different for traditional and financial companies, involves the crucial role of the board of directors with different effects on the business and highlights how environmental issues are also prominent in top management skills (Soana, 2024).

Many companies carry out "greenwashing" strategies, i.e. cover-up activities aimed at building a certain type of reputation or making public opinion forget about unpleasant or less-than-virtuous practices, declaring that they are responsible. These practices, banned by the European Parliament in 2024, which intervened to protect consumers' informed purchasing decisions towards more sustainable models, rarely start from the product's design, materials, and production techniques. Research into the materials used, whether recycled or sourced from supply chains, gives crucial direction right from the initial stage of the product life cycle, which goes as far as considering product renewal and durability.

A process that today cannot fail also to consider the "end of life" and how the product will be disposed of and possibly reused. In a nutshell, it is a matter of changing the existing paradigm and shifting the search for value into services revolving around the product, which may be of a different nature: preventive maintenance, warranty extension, subsidised financing, training, etc.

Another key theme is the reporting of corporate activities of a non-financial nature (Soana, 2011), such as decarbonisation, which requires a transparent narrative to avoid the paradox that sees multinational companies with extensive use of fossil fuels transforming, in a short time, into "Carbon Neutrality" companies. In this respect, the European Union has envisaged the revision of the emissions trading system (ETS), the revision of the Market Stability Reserve, the implementation of the carbon leakage tool, the sharing of efforts between EU countries, carbon leakage, a guarantee fund for a fair energy transition, strengthening rules to increase carbon removal in the land use sector (LULICF), a review of transport emission quotas aeroplane, etc.[2]

Therefore, making production processes more efficient takes on a different meaning from producing at the lowest possible cost. However, it broadens the reasoning on all phases of the value chain from which a marketing positioning based on the real values sought and proposed to customers should come.

Enterprises of every size know the importance of reputation in the economic system. Some are born with a "circular footprint". For example, using waste that is transformed into products can create value right from the beginning of their history. Instead, others transform over time for competitive and survival reasons. This transformation is sometimes only apparent and generates externalities whose impact is measurable only after years.

Unfortunately, the paradox for some raw materials, such as plastic, is that recycled ones have a higher cost than virgin ones, preventing their reuse and preventing the continued offering of similar products whose unit cost is very competitive.

In this scenario, where 80% of the plastic found in the oceans comes from around a thousand rivers, there are plenty of initiatives worthy of being reported and which have already led to important results, such as the direct removal of tons of plastic. However, the problem will not be solved until the local communities living along the rivers continue to apply these incorrect disposal practices and until a solution to the use of disposable plastic is found. A scenario in which companies that produce and employ millions of jobs worldwide collide and are not yet capable of developing a functioning and low-cost transition, as in the electricity sector.

Italy is one of the countries with the highest recycling percentage in Europe and is well above the average in the old continent (Fondazione per lo sviluppo sostenibile, 2023). This data does not mean a circular economy has been introduced into the economic system. However, it highlights the effort made in recent decades and how much work still needs to be done to guarantee real circularity based on continuity. Certainly, historical reasons have contributed to this practice, such as the practice of recovering and passing on to new generations the immense cultural heritage with which the country is endowed, but also eating habits based on quality products whose obtaining requires safeguarding the territory that becomes a "food and wine basin" based on small productions of excellence. In conclusion, the second largest manufacturing company in Europe, after the German one, is a processing company whose success depends on the product's design, efficiency, and the fight against waste of raw materials which are necessarily imported.

[2] https://www.europarl.europa.eu/topics/it/article/20200618STO81513/green-deal-europeo-la-chiave-per-un-ue-sostenibile-e-climaticamente-neutrale

9.4 Hypotheses Development

The '2030 Agenda' has often been referred to as a fundamental goal for the development of the circular economy. However, since its signing in 2015, pandemics, international crises and wars have demonstrated how global complexity can slow down its implementation, making some choices questionable, such as the transition to electric.

Certainly, the end of the Fordist model has imposed a change of mentality and vision in the business world, especially in Western countries. It is often the leading companies that are willing to change and invest in medium- and long-term projects. However, one company alone is not enough to save the ecosystem if others engage in improper practices. This is why networking with institutions and territories (Gulick & Urwick, 1937), where the degree of corruption in the system is at its lowest, is essential to ensure that the product is not something taken out of context but has a direct positive effect on populations and the environment.

Starting from the company vision and the consequent product design phase, it is necessary to consider the production processes, the distribution model, the enabling factors, the 'end of life of the product' and the people to be involved, including the stakeholders (Freeman & Evan, 1993). The 'end-of-life of the product' will be crucial in the circular economy because it can return value to be reused for further manufacturing choices. To achieve this, the company must make significant investments in research, development, and experimentation. This will enable the company to deeply understand the production and value generation processes, considering the raw material as a material and corporate asset.

Therefore, the product's design becomes crucial to activating a modular construction process that still allows the replacement of parts, repair, and reuse. It's about resetting a logic that is contrary to disposable products, which apparently cost less but can have devastating impacts on the environment. This choice clashes with years of bad habits and wrong practices, such as the waste resulting from the catering sector. Especially in rich countries, the "Doggy bag", which has become mandatory in some European states, is entering the Western consumer mentality for ethical and educational reasons after overcoming limits linked to customs and compromising social status. Considering that a third of restaurant food turns into rubbish, wasting value and energy, it is easy to understand how important the educational aspect is towards ethical and environmental issues and the regulatory aspect for establishing and applying good practices among the population. In the latter area, governments clash with resistance to change from industrial lobbies and small entrepreneurs living off the "traditional economy". This resistance is due to the fear of losing profits, exiting the market, and not being able to survive.

In order to do this, public communication will have to be transformed into educational information and not mere marketing. It will be effective when it brings the population, economic operators and territories closer to environmental problems, explaining how to solve them and demonstrating that it is impartial and transparent. This last point is not trivial if you consider that very often public, or even state-owned, companies have been the main source of pollution in an area and the main cause of illnesses among the resident population.

Starting from the vision of the entire value chain in product development, in addition to the importance of design, right from the research phase, it is necessary to remember the advent of "Servitization", that is, the transition from the logic of only selling the product to that of an overall service that often includes extended warranty, scheduled maintenance, field assistance, replacement and management of used equipment thanks to maintaining the value of the product, etc. An approach that clashes with the strategy linked to "cost leadership", perceived in some areas of the globe and by some groups of consumers and businesses, which requires purchasing a product at the lowest possible price, without considering the TCO, the LCC[3] and construction features.[4] The transition towards "Servitization" requires educating and training consumers and enterprises towards an expanded vision of the topic and directing them towards conscious long-term economic choices.

This expanded vision is described, for example, by the 'Food system', through which awareness of the interconnectedness of systems and processes is raised. A step that needs special attention is "Servitization", which some companies see only as a way to improve turnover, margins and customer loyalty. From this point of view, the sale of services could compensate for the decline of saturated markets and offer companies business opportunities.

Therefore, this logic applied to food no longer considers it only as food for its own sake, but also as nutrition, health, rural development, sustainable agriculture, waste reduction, pollution control, repopulation, species maintenance, etc. Good food will have to take all these issues into account because the failure of just one of them could lead, in the long run, to the failure of the entire value chain.

From the point of view of the business community, it is vital to understand how ready businesses are for the circular economy through an approach that starts with the sourcing of raw materials and ends with the distribution network and, above all, the return network.

A transition to circularity must include an internal structure that goes beyond the product and involves the Stakeholders (Sacconi, 2004). At an international level, some models, such as the Life Cycle Assessment (LCA), have established themselves[5] which, for the past fifty years, has allowed us to quantify the impacts of a product on the environment and human health by identifying the critical phases and the areas on which to reduce the environmental footprint. LCA is also crucial for developing environmental labelling schemes such as the Environmental Product Declaration (EPD), through which the environmental impacts linked to producing a specific quantity of product or service are described.

---

[3] Life-Cycle Costing.

[4] Total Cost of Ownership.

[5] This methodology is standardized internationally through the ISO 14040 and ISO 14044 standards.

9.5 Conclusions

This is an issue that is also becoming geopolitical, given the different positions of various governments in pursuing sustainability and waste reduction policies. As in the case of CSR, there are governments that are investing heavily in the reuse and remanufacturing of products and others that are struggling with waste management, which have become true ecological bombs. The very choice to dispose of waste in other countries, which are able to propose appropriate and functioning technologies and solutions, changes the relationship between governments and proposes projections of influence that can be decisive in the near future. The transition to electric in automotive itself is generating significant issues in relation to battery disposal and the sustainability of the business, given the use of raw materials for construction that heavily affects the planet. This paper has analyzed the geopolitical and geoeconomic impacts of circular economy and how it can become a force in the foreign policy of nation states.

The circular economy model discussed in this work is essentially simple to understand and, above all, to share from an elementary and natural level of cognition. Individuals and families around the world can apply it. The education and application of the circular economy should start precisely in the basic cell of society, which means preserving territories and, consequently, memory.

Implementing the circular economy still requires much work because new needs and technologies impose new challenges. For example, electronic waste, which is difficult to recover and prevalent with the increase in IoT-related devices, requires experimentation with new recycling and reusing logic despite ICT manufacturers' tendency to produce closed and no longer repairable products.

A single state, economic operator and industrial sector cannot easily resolve the ongoing challenge. The complexity is such that the different economic systems, from local to global, already interconnected by financial platforms that sort payment flows, should learn to continuously adapt to economic and financial changes and climate (Rifkin, 2014), social and political changes. The change, however, must not be a passive resolution of the problem in order to contain losses or maintain profits, but to address why the problem arose and how to solve it with new methodologies that take a holistic view of the phenomenon, including changing consumer preferences.

In the future scenario, the increase in the global population will heavily impact energy sources, agriculture, global food supply chains, etc. Paradoxically, tools designed for the success of globalization, such as open international markets and long supply chains, have already proven to be unsuitable for solving problems due to the costs of maintaining them and the lack of resilience of the natural system. Allowing consumers thousands of kilometres away to consume an out-of-season product that has become fashionable requires reconsidering the entire cost of the supply chain and asking whether it would not be cheaper and healthier to consume local products without them becoming a drain on the system. It is about reconsidering the economic logic of businesses and consumers, primarily through education, considering biodiversity and the resilience of the ecosystem.

This problem also concerns certain consumerist and exaggerated lifestyles, often imitated by developing countries whose social rise is, paradoxically, confirmed by access to this type of consumption. For example, the packaging of single portions created for singles, which has become a marketing objective of distribution companies and beyond, has generated an enormous amount of waste that wastes resources and energy and could be reduced with simple measures and more responsible behaviour.

In this complex global scenario, we must be careful not to create a false circular economy to renew consumption, create new needs and gain market share. For example, the issue of electric cars, for which the EU had set a date for the transition to electric cars in 2035, is highlighting major problems in the construction of batteries, their safety, their autonomy, and their disposal. Many today wonder whether this technology, for which raw materials such as lithium are needed, is truly beneficial for humanity or just a technology brought forward by some states for geopolitical reasons aimed at obtaining leadership and global influence to the detriment of others. Some questions arise spontaneously. Are there alternatives to electric power, and why have these yet to be considered by European policymakers? Could biofuels generated from waste and crop waste such as soy, corn, wheat, chard and palm oil once again become the solution to be applied in the automotive sector? Some new groups are moving along this path in which the time factor will still be crucial to allow the technology a timely "go-to-market". A path which, however, could clash with the great geopolitical interests linked to the "rare earths" and their possession and use. Dynamics that could drastically change global balances, relegating sustainability and the circular economy to the background. These evolutions see the extension of the concept of national security to issues considered "without a passport", such as global food security, for which, in the USA, constant state supervision is required (Drezner, 2024). Expanding the boundaries of national security, formerly limited to a few specific dossiers, could see its effectiveness diluted as each potential threat requires multiple degrees of response and the participation of the international community, on which the pressure of global powers will be greater. Moreover, entrusting these issues to new bodies could generate additional bureaucracy with very high costs for citizens and results that are not easy to quantify. Paradoxically, even liberal democracies' imposition on developing countries of the recognition of some social priorities, particularly felt by a minority part of the Western population, could generate ideological neocolonialism (Benedetto, 2009) far from the real circular economy and the benefits that it could produce Pontificio Consiglio della Giustizia e della Pace (2004).

Circular Economy requires a change of mentality on the part of all players and the search for a "balance of the system" that, overcoming the traditional logic of partisan objectives, directs the effort towards the common good (Franchi, 2024). How to reconcile the survival of a rural community living in the mountains with the one of the shareholders of a large company listed (Zamagni, 2020) on the stock exchange and governed by the logic of 'shareholder value' will be the challenge of the near future and of new generations.

References

Benedetto, X. V. I. (2009). *Lettera Enciclica Caritas in Veritate*. Libreria Editrice Vaticana.

Braungart, M., & McDonough, W. (2002). *Cradle to cradle. Remaking the way we make things*. Nort Point P.

Bruni, L. (2015). *Il mercato e il dono, gli spiriti del capitalismo*. Università Bocconi Editore.

Chomsky, N. (2017). *Le Dieci Leggi del Potere*. Ponte alle Grazie.

Coda, V., Invernizzi, G., & Russo, P. (2001). *La strategia aziendale*. McGraw-Hill Education.

Drucker, P. (2000). *Manuale di Management, compiti, responsabilità e metodi*. Etas.

Fassio, F., & Tecco, N. (2018). *Circular economy for food*. Edizioni Ambiente.

Francesco. (2015). *Lettera Enciclica Laudato Si', sulla Cura della Casa Comune*. Libreria Editrice Vaticana.

Franchi, M., & Rainieri, B. (2019). *Riflessioni sul Management Responsabile*. Licosia Edizioni. https://hdl.handle.net/11380/1282557

Franchi, M. (2024). A Critical Reflection on Corporate Social Responsibility: Between Communitarian Ethics and Business Purpose. In Chapter in the book Economic systems and human rights: Using socioeconomic models and practices to promote global economic socialization. Palgrave Macmillan. https://hdl.handle.net/11380/1354566

Freeman, R. E., & Evan, W. M. (1993). In T. B. e. N. Bowie (Ed.), *A stakeholder theory of the modern corporation: A Kantian capitalism*. Ethical Theory and Business.

Friedman, M. (1987). *Capitalism and freedom*. Studio Tesi. (original work published University of Chicago Press, Chicago 1962).

Gulick, L., & Urwick, L. (1937). *Papers on the science of administration, Institute of Public Administration*. Columbia University.

Kim, W. C., & Mauborgne, R. (2005). *Strategia Oceano Blu*. Harvard Business School Press.

MacArthur, E. (2024) *Navigating the circular economy reporting landscape*.

Malgorzata Jakim, W., Samokhalov, V., & Baldassarre, B. (2024). Achieving European Union strategic autonomy: circularity in critical raw materials value chains. *International Affairs, 100*(4), 1735–1748. https://doi.org/10.1093/ia/iiae127. Published by Oxford University Press on behalf of The Royal Institute of International Affairs.

Molteni, M., & Alessandra, T. (2008). *Responsabilità Sociale d'Impresa*. IlSole24Ore.

Rainieri, B. (2019). *Appunti di psicologia: Spunti etici per i leader globali*. Parma, Italia: s.e.

Rifkin, J. (2014). *La terza rivoluzione industriale*. Mondadori.

Santesso, E., & Sòstero, U. (2004). *Principi contabili per il bilancio d'esercizio*. IlSole24Ore.

Soana, G. (2024). Environmental strategies, environmental performance and board sustainability committees: Are financial and non-financial companies different? *Research in International Business and Finance, 69*(9), 102208.

Sombart, W. (1978). *Il borghese. Lo sviluppo e le fonti dello spirito capitalistico*. Longanesi.

Stahel, W. R. (2019). *Economia Circolare per Tutti. Concetti base per cittadini politici ed imprese*. Edizioni Ambiente.

Stiglitz, J. E. (1997). *Il ruolo economico dello Stato*. Il Mulino.

Vandana, S. (2023). *La Vita è Maestra*. Piemme.

Weber, M. (2010). *L'Etica Protestante e Lo Spirito del Capitalismo*. RCS Libri.

Wookey, C., Alford, H., & Hickey, L. (2018). *Advancing the common good through purpose-led business: Early learning from a blueprint for better Business's work with some major companies*. Blueprint for Better Business.

Zamagni, S. (2020). *Disuguali*. Aboca Edizioni.

Zingales, L. (2012). *Manifesto capitalista*. Rizzoli.

Magazines Articles

Drezner, D.W. (12 August 2024). "*How everything became National Security.*" Foreign Affairs. https://www.foreignaffairs.com/united-states/how-everything-became-national-security-drezner
Franchi, M. (2022a, Giugno). La Responsabilità Sociale d'Impresa tra Moda ed Etica. *Oikonomia-Journal of Ethics & Social Sciences,* N2, 28–31. https://hdl.handle.net/11380/1281219
Franchi, M. (2022b, Ottobre). Scopo dell'Impresa ed Etica Manageriale. Oikonomia-Journal of Ethics & Social Sciences, N3, 37–39. https://hdl.handle.net/11380/1288626
Sacconi, L. (2004, supplemento febbraio). Responsabilità Sociale come governance allargata d'impresa: un'interpretazione basata sulla teoria del contratto sociale della reputazione. Liuc Papers n., 143, 1. Serie Etica, Diritto ed Economia 11.
Soana, M. G. (2011). Corporate social responsibility and financial performance: Evidence from the financial sector. *Corporate Ownership and Control, 8*, 27–36.

Report

Blueprint for Better Business. (2018a). *How can investors identify purpose-led companies.* Blueprint for Better Business.
Blueprint for Better Business. (2018b). *The what, the why and the how of purpose.* Blueprint for Better Business.
Blueprint for Better Business. (2019). *Fairness in business.* Blueprint for Better Business.
Blueprint for Better Business. (2020). *Purpose for PLCs—Time for Boards to focus.* Blueprint for Better Business.
Concilio Ecumenico Vaticano II (1965). *Costituzione Pastorale Gaudium et spes sulla Chiesa nel mondo contemporaneo.*
European Commission. (2023). *Commission Staff working document: impact assessment report.* Accompanying the document 'Proposal for a regulation of the European Parliament and of the Council establishing a framework for ensuring a secure and sustainable supply of critical raw materials and amending Regulations (EU) 168/2013, (EU) 2018/858, 2018/1724 and (EU) 2019/1020', SWD (2023) 161 final (Brussels: European Commission, 2023), https://eur-lex.europa.eu/legalcontent/EN/TXT/?uri=CELEX%3A52023SC0161, p. 11
European Commission, The European Green Deal, & COM. (2019). *640 final.* European Commission. https://eur-lex.europa.eu/resource.html?uri=cellar:b828d165-1c22-11ea-8c1f-01aa75ed71a1.0002.02/DOC_1&format=PDF
European Commission. (2021). Updating the 2020 new industrial strategy: building a stronger single market for Europe's recovery, COM. In *350 final.* European Commission. https://eur-lex.europa.eu/legal-content/EN/ALL/?uri=CELEX%3A52021DC0350
Fondazione per lo sviluppo sostenibile. (2023). *Rapporto 2023 condotto dalla con il patrocinio del Ministero dell'Ambiente e della sicurezza energetica e di Ispra.*
Pontificio Consiglio della Giustizia e della Pace. (2004). *Compendio della Dottrina Sociale della Chiesa.* Libreria Editrice Vaticana.
Regulation (EU). (2024). *2024/1252 of the European Parliament and of the Council of 11 April 2024 establishing a framework for ensuring a secure and sustainable supply of critical raw materials and amending regulations (EU) No 168/2013, (EU) 2018/858 and (EU) 2019/1020 (known as the Critical Raw Materials Act or CRM Act) (Brussels: Official Journal of the European Union)*, 3 May 2024, https://eur-lex.europa.eu/legal-content/EN/TXT/?uri=CELEX%3A32024R1252&qid=1716542362110. It is worth noting that this final version of the law increased the recycling rate from the 15% proposed by the European Commission in the proposal of March 2023.

The Royal Institute of International Affairs, & Chatham House. (2019). *An inclusive circular economy priorities for developing countries*. Felix Preston, Johanna Lehne and Laura Wellesley Energy, Environment and Resources Department.

Massimo Franchi is a strategic consultant for multinational organisations and Senior Research Fellow at the University of Modena and Reggio Emilia in the Department of Law and lecturer at the Universidad Nacional de La Plata—Unesco Catedra. At Parma University he is a member of the Laboratory of Law of Markets and New Technologies (DiMeTech) and has been junior and senior research fellow, lecturer and adjunct professor in the integrative course of Political Risk of the FRIM master's degree. At the University of Calabria, he was adjunct professor of the II Level University Master Course in Intelligence and member of the Intelligence Lab. He is a lecturer at the ITS Academy Tech&Food and research director of the FMS 'Winter School of Geopolitics".

Chapter 10
Beyond Sustainability: Circular Economy and Integral Development A Problem of Economic Anthropology

Francesco Poggi

Abstract The simple equivalence between sustainability and the circular model risks distancing the real solution to the many problems that beset the various economies. Certainly, it may favour the improvement of certain parameters and careful management of processes, from the supply of raw materials to the reduction of waste and emissions, from the rationalisation of methods to the reduction of waste, finally placing a limit on the exploitation of resources. But this may not be enough. Indeed, the revision of the many pre-analytical postulates that guide the choices and practices of the real economy can no longer be postponed; the transformation of the development model can no longer be postponed either. For we are faced with a question of economic anthropology. The circular model must therefore be accompanied by an integral approach to the economy that encompasses all aspects, from cultural to ethical, from demographic to environmental. In search of an integral ecology, of an integral development that puts the person, the environment and the responsible enterprise at the centre, no longer focused only on business, but oriented towards the territory and relationships. Not only because it is convenient, but because it is in its very nature and mission. Only in this way can the circular economy produce the great effects expected. A vision of the world and not a mere palliative.

Keywords Circular · Sustainability · Integral ecology · Development · ESG · 2030 agenda · Recycle · Reuse · Regenerate · Eco-design · Economic anthropology · Waste · Smart cities · Third sector and volunteering · Mobility · Artificial intelligence · Forest and natural cycles · Natural communities

F. Poggi (✉)
Pisa University, Pisa, Italy
e-mail: francesco.poggi@ec.unipi.it

S. Patnaik et al. (eds.), *Circular Economy and Sustainable Value Creation through Eco-Innovation*, Circular Economy and Sustainability,
https://doi.org/10.1007/978-3-031-88516-7_10

10.1 Foreword

For years, formal rationality has been the compass for decisions in the economic world. On the basis of it, only choices along the lines of logic, probability calculus and utility maximisation have been accepted. A real monopoly of the formal aspects, to quote Polanyi, stifling any alternative approach. But it is also everyday practice to have an economic policy in which the econometric forecasts of central banks, governments, and research centres on inflation, growth, and unemployment are not systematically confirmed, or expectations on consumer behaviour are disregarded. In this uncertain and variable scenario, there are the current practices of resource exploitation which, added to the constant upward trend in world demography and the ever-evolving climate crisis, pose serious questions about the resilience, in the future, of economic systems as well as the social cohesion within individual states. In fact, while the great race for technological innovation, accompanied by huge public and private investments, is geared towards overcoming various structural criticalities (e.g. the ever-increasing energy supply needed), it in turn leads to new criticalities. The economic development paradigm has proved incapable of resolving the many issues of the contemporary global economy. Introduced in a phase characterised by the need for post-conflict and post-colonial reconstruction, it has been oriented exclusively towards the growth of individual countries' economic performance, with a focus on GDP. The very policies of so-called 'underdeveloped' or 'developing' countries were modelled on this basis. One example of this was the role, the mission, of the International Monetary Fund, established in 1944, in planning investment policies. With a methodology guided from above, it became universal, but abstract and homologising. Little importance has been given to local, geographical, environmental, cultural, or civil characteristics in economic policy choices. The weaker and backward areas have over time become the theatre of exploitation for the procurement of raw materials, while the more developed parts have become immense consumer and business markets. The great boom in quantitative and qualitative growth is undeniable, with clear progress in terms of, for example, improved living conditions, a higher average life expectancy, and access to health services. At planetary level. But this growth has not been balanced by an equally significant human, social and civil development, in the face of enormous environmental devastation and significant climate change. The panorama, at the beginning of the third millennium, has also seen growing tensions between areas of the world, the explosion of world debt (around 315 trillion dollars, 2024 figure) and new financial crises, with a further widening of inequality, both internal and external, and an increase in relative poverty. The mirror of this scenario are the large cities, the ever more imposing metropolises, which present their bill in terms of mobility, logistics, services, liveability for the elderly and children, pollution, migration and coexistence, which are increasingly difficult. To all this, both academic economic theory and international economic policy have responded by including the concept of sustainability in their programmes and objectives. According to this more holistic approach, human well-being, social performance and, at least in intentions, the management of the most pressing environmental issues should be improved. In this sense, all the forecasts for action contained in the UN 2030 Agenda, such as the commitments made at the annual international climate conferences (from 1992 in Rio to Dubai 2023) should be read. In this sense, both the ESG criteria (Environment, Social and Governance) that, with the 17 goals, are progressively entering into the strategic lines of large companies and public bodies, and the sustainability report, a real tool for monitoring, reporting and communicating the new management and planning process, which must be precisely responsible, must be interpreted in this sense. Investing in ESG therefore means directing capital towards companies that respect the environment, are attentive to social inclusion and rights, between containing poverty and inequality and improving key environmental parameters. Around these themes, however, some fundamental, systemic questions are also maturing, to which the entire scientific community will be called upon to respond. First of all, whether the market economy is the paradigm on which to focus to achieve these global goals. The ESG criteria are in fact to be inserted into a theoretical-practical structure that would keep unchanged those general models that refer to a very precise economic philosophy, according to what Schumpeter called the pre-analytical vision, not scientific but ideological, by definition, expressed in that neoclassical 'toolbox', the true and proper supporting skeleton of contemporary economic systems. Thus, if the concept of sustainability implies an attempt to 'save' the basic theoretical model, avoiding throwing all the economics textbooks to the wind (or rather to the rubble) or questioning the very training of managers as well as public planners, it might not be decisive, since it does not question the system's load-bearing pillars, starting with why economics is done, for what and for whom, for what ends or objectives. This scenario would simply be geared towards making the entire framework of the Market Economy sustainable. And to achieve this we are all going to school in sustainability. Initially, it was nothing more than a task and a set of activities in the hands of certain professional figures in different companies, such as the corporate social responsibility manager or the sustainability manager. In recent years, it has spread to all corporate functions and all areas of a company's operations, from strategy to marketing, from operations to supply chain and innovation. Sustainability to be valued as a lens to filter all business processes and above all as a new and decisive competitive factor. Even business schools and universities are now called upon to train new skills, with specific courses on ESG criteria in finance, sustainability reporting, accounting measurement systems, the concept of purpose, and innovation for impact. Among the decisive tools for fully affirming the concept of sustainability, that of the circular economy is emerging with increasing conviction and diffusion, which entails reviewing, both upstream and downstream, production processes and resource management. From energy to logistics, from the waste cycle to ecodesign, the adoption of a circular model will have to lead us to reduce waste, optimise processes and save time. It is a necessity that can no longer be postponed, no longer an extravagant or at best niche idea, as it was presented in the recent past. But the simple equivalence between sustainability and the circular model risks distancing the real solution to the many problems that beset the various economies. Certainly, it may favour the improvement of certain parameters and careful management of processes, from the supply of raw materials to the reduction of waste and emissions, from the rationalisation of methods to the reduction of waste, finally placing a limit on the exploitation of resources. But this may not be enough. Indeed, the revision of the many pre-analytical postulates that guide the choices and practices of the real economy can no longer be postponed; the transformation of the development model can no longer be postponed either. For we are faced with a question of economic anthropology. This discipline is concerned with the link between production mechanisms and the allocation of economic resources (not only physical resources but also intangible assets such as power, *status* and social prestige) and the socio-cultural environment of a given community. Indeed, there are not two separate crises, one environmental and one social, but a single, complex socio-environmental crisis. The circular model must therefore be accompanied by an integral approach to the economy that encompasses all aspects, from cultural to ethical, from demographic to religious. An integral development that is capable of bringing together environmental phenomena and problems (global warming, pollution, depletion of reserves, deforestation, etc.) with issues that are not normally associated with the ecological agenda in the strict sense, such as liveability or the beauty of urban spaces. The integral approach goes precisely in the direction of combating poverty, restoring dignity to the excluded and caring for nature, refusing to respond to problems with a series of urgent and partial measures. Technology itself is unable to see and assess the multiplicity of relationships between man and man, between man and the environment and between man and things. The anthropological problem manifests itself in the excess of anthropocentrism that undermines any attempt to strengthen social ties, as in the mannerist ecologism that renounces taking a systemic view. That is why sustainability is not enough, that is why the circular model will have to be guided with an integral approach, involving and enhancing the different cultural riches of peoples, Art, the third sector, voluntary work and gratuitousness, the idea of chorality and cooperation, a new project of coexistence, of mobility (smart city), the revaluation of natural riches, natural cycles and natural communities. In search of meaning. On the one hand, the paradigm of Industry 4.0., with artificial intelligence, robotics, accompanied by eco-innovation and circle economy. On the other, a social transformation that enables people in a given local context to live in accordance with what they feel close to, in accordance with what they value most, and to achieve what they most aspire to do and/or be. Social value transformation will therefore be achieved by integrating knowledge, eco-innovations, human resources, with technology, in a harmonious blend of nature, people and industry. In search of an integral ecology, of an integral development that puts the person, the environment and the responsible enterprise at the centre, no longer focused only on business, but oriented towards the territory and relationships. Not only because it is convenient, but because it is in its very nature and mission. Only in this way can the circular economy produce the great effects expected. A vision of the world and not a mere palliative.

10.2 The Circular Model: The Basic Approach

The circular economy model aims to propose a regenerative economic system in the long term in contrast to the current dissipative and consumerist approach of business-as-usual. It tends to preserve natural capital through reduced drawdown, to optimise the performance of resources through their reuse, regeneration and maintenance, and finally to identify and eliminate any negative externalities (Georgescu Roegen, 1976). According to Bompan (2016), the paradigm of the Circular Economy is based on an economic model that satisfies three fundamental principles: rediscovering the discarded old material as a source of a new one; no material represents an unused value; avoid premature death of materials. The thought goes back to agricultural economies before the industrial revolution, where everything that could be repurposed was reused or recycled: women used old clothes to recreate valuables for the home; with olive oil they made soap and with wine they made vinegar; with wood they made coal for heating and the fields were fertilised with animal waste. Later, an economic system based on immediate exploitation was implemented, and goods had a shortened life cycle that became a linear cycle. Today's economy is in fact based on the linear approach: 'take, make, use, dispose'. It is called linear because once consumption ends, the product cycle also ends and it becomes waste. Products are designed to meet a single need and diversification seems more important than the need itself. As a result, products are bought, used and thrown away again and again, instead of being repaired or reused. In the natural ecosystem, on the other hand, there are no landfills. Everything that is waste for one species is food for another species. The sun provides energy, things grow, die and provide the earth with nutrients. The cycle begins again each time. This natural model has worked perfectly for millennia and it is precisely this that inspires the circular economy, obviously updated to the current technical and civilisation stage. Thus, both an individual and a company that wants to remain competitive can make the best use of resources, putting them back into the production cycle instead of dumping them in landfills and turning them into waste. To do this, we will have to design products in such a way that, after using them, we can return the materials from which they are made to those who produced them and return the biological parts to the environment. Added to this is the assumption that these products are created and transported using only renewable and clean energy. As far as biological waste is concerned, this involves redesigning the products into their parts and packaging, using safe and compostable materials. In the case of technical materials, which are not made from biodegradable materials, recovery involves recycling metals, polymers and alloys so that they continue to retain their qualities and can still be useful beyond their original intended use. In this way, today's products could become tomorrow's resources, a closed-loop model in which 'waste equals food'. The term 'circular' denotes precisely this continuous flow with a system based on 'take, make, use, return'. This strategy envisages a shift from 'cradle-to-grave' to 'cradle-to-cradle' waste management with an ever-increasing decrease in waste production through eco-innovation and the use of new business models. The so-called Green Economy, which is based on the efficient use of resources, the clean and safe production of goods and the reduction of pollution, is one facet of this. In the contemporary circular economy, concepts such as eco-design, repair, reuse, maintenance, product sharing, remanufacturing, prevention of waste production or zero waste and recycling are becoming increasingly popular. All stages of production and thus all different applications are under discussion. Starting upstream from design and conception with eco-design. There is a tendency to change the way products are designed, considering from the outset their possible end-of-life use through features that allow them to be disassembled or refurbished. In this sense, it is possible to design products or parts of them (design for components) to be repaired and remanufactured later (Bompan & Brambilla, 2016). With appropriate design or, better still, with ecodesign, true innovation in a circular sense starts right from the conception, design and development phase of a product or service. Also shaping possible market scenarios. It is therefore essential to rely on Life Cycle Thinking approaches, i.e., to carry out analyses, applying methodologies that take into account the impacts generated during the entire life cycle of the product, starting at the very end of its life cycle. Eco-design envisages that its various components can be easily disassembled to be recovered and reused, including the types of materials used. With this type of design, each product must be conceived as modular to allow the replacement of parts, recovery and reuse, both of the product and its components taken individually. Repairability and maintenance are also an essential part of the design, as the product must be able to be repaired by replacing technologically obsolete or damaged parts and must be able to be maintained in such a way as to allow its life cycle to be extended (Barbero, 2012). Hazardous, toxic or highly polluting substances must be replaced with others that do not contain harmful substances to make products more easily recyclable. When judging a requirement such as reparability, different results can be obtained depending on the methodology chosen for the measurement. Due to their longer lifespan, repairing products can, in principle, reduce resource consumption in terms of raw materials and energy, and thus the volume of waste and greenhouse gas emissions caused by them. On the one hand, eco-design standards require that the design of products facilitates their reparability, while on the other hand, action is taken to reduce other barriers to repair. But beware, repair is not always automatically the most ecologically sustainable option. If, for example, a new refrigerator consumes much less electricity than the old repaired appliance, repair is not the most environmentally friendly option. The next step is to extend the life of the product, trying to give modularity, versatility and adaptability to the products so that they can be adapted to changing external conditions and recurring maintenance to extend the life of the product as much as possible. Then, we will have to rely more and more on renewable energies, trying to gradually abandon the old model based on fossil energy sources, such as material recovery, preserving quality through recycling, reuse, regeneration, to gradually replace them with virgin raw materials. Again, we will have to create sharing platforms that are based on the standard management of products among users themselves to increase the utilisation of goods and at the same time decrease the production of new, unnecessary goods. Sharing also reduces the cost of access to products and services and creates a situation of social interaction and cohesion. But the guiding idea is the product as a service, through the sale to the customer of benefits associated with the product itself. This opportunity has taken hold especially with the advent of new digital technologies and leads to a more meaningful interaction between customer and company. Finally, the focus on end-of-life will become increasingly strategic. Products that reach the end of their life can be transformed, remanufactured and reintroduced into the production cycle generating a higher value than their initial destination, and this process is called 'upcycling'. On the other hand, open-chain recycling uses recovered materials to generate lower-value products, called 'Downcycling'. For example, biorefining extracts small quantities of valuable materials or transforms waste into energy. The increase in related services (maintenance and repair), should lead to increased employment. Even if its diffusion remains rather small, its gradual strengthening in logistics processes and supply chains will be able in time to ensure, increasing benefits in environmental, economic and social terms. In order for the transition to the circular economy to be successful, it is essential to support and develop new technologies that improve supply chains, reduce the environmental impact of companies, and make it possible to minimise the share of resources that is wasted during the production and use of products. The study of risk management practices in circular supply chains is also still at an early stage and needs to be further investigated, as it is characterised by a higher degree of uncertainty than the traditional one, having to involve two different material flows, upstream and downstream. Therefore, it is necessary to create an institutional scenario favourable to the formation of new circular enterprises or to incentivise the abandonment of the linear model, characterised by a strong dissipation of value during its entire life cycle. The main actions to be taken to encourage circularity can be: an intensification of incentive measures from a financial point of view, or tax relief, for companies that operate sustainably; a strengthening of the taxation on the production of waste; the creation of infrastructures and information systems that can facilitate the flow of resources; the introduction of recognisable and global applied labelling specifies information regarding the history and characteristics of the materials used. The path is now marked. However, although the circular economy has many advantages, there is still a huge gap between its theory and its implementation. Many companies have started to implement circular economy practices and to develop circular business models, but they face several obstacles of philosophy, market, organisation, system. Moreover, not everyone is aware of the potential opportunities that the circular economy presents, also in terms of value creation. For companies, in fact, this shift could represent a potential source of non-negligible competitive advantage. In addition, good and careful management of processes, such as the supply of raw materials used, could allow a reduction in waste and emissions that can make a difference both from an economic point of view, positively impacting costs, and from that of corporate image. For the economy as a whole, the shift to the circular model seems obligatory to reverse certain long-term trends that are difficult to sustain. But for the result to be at its best, a systemic approach is needed, thinking of the whole as well as the individual parts of which the process is composed, taking into account the cause-effect relationships between the various components. From the basic approach we must move on to the integral approach, questioning the very logic of doing business, touching every sphere of civil and economic life. From the ecological transition to the digital transition, from the social transition to the urban transition. The city, with the spheres of urban mobility, logistics (reverse logistics) and urban freight (smart city); labour, rights, democracy, competition and coexistence; energy, its production and diffusion; technology, with the new frontiers of digitalisation, the Internet of Thinghs and artificial intelligence, robotics and Big Data; the widening of the economic perimeter in the area of volunteering and gratuitousness (third sector and cooperation); ethical and green finance, private and public investments; the management model of public bodies themselves, corporate, commercial and market regulations; the redevelopment of the management of farms, fish farms and livestock breeding; lastly, the management of natural cycles (forest, water) and the great theme of natural communities. A radical change in terms of economic anthropology, philosophy, motivations and objectives. Beyond sustainability.

10.3 The Circular Model and the 2030 Agenda: An Integral Approach

Why do so many species invest a significant part of their energies in helping the weaker individuals within their communities? This question has recently arisen not in the philosophical or moral sphere but in the botanical one. Perhaps unexpectedly. Thus, it turns out that in natural communities, mutual support is the most efficient system to ensure the survival of all (Mancuso, 2023). In parallel, the great debate on artificial intelligence, Big Data, and robotics brings two protagonists of world industry such as Sam Altman, CEO of Open AI, and John Elkann, CEO of Exor and chairman of Stellantis and Ferrari, to discuss innovation and community (Italian Tech Week, Turin, September 2024). The proposed recipe is always the same: transition and sustainability. The term 'transition' first appeared during the 1970s, in particular in the first report of the Club of Rome (Meadows, 1972), which insisted on the need for a 'transition from a growth model to one of global equilibrium', emphasising the ecological risks induced by economic and demographic growth (Bischoff, 2017). Also in the 1970s, people began to talk about sustainability. Economic, through the concepts of efficiency and growth, of the relationship between public resources employed and results in terms of quantity and quality of service. Social, through access to territorial opportunities, responses to interaction needs, participation and social inclusion (access in physical, monetary and quality terms, in the sense of less urban congestion and better urban spaces). Environmental, through the reduction of the negative environmental effects of human activity related to travel, the reduction of harmful emissions, the reduction of noise pollution. Thus development becomes a process of change such that the exploitation of resources, the direction of investments, the orientation of technology and institutional changes are made consistent with future needs as well as current ones. What is missing from all this, however, is the overall vision, the awareness of how everything is linked, connected. The integral approach. This is the concept of interdependence, which is also central to quantum physics and the new cosmology. The circular economy takes on meaning and can become decisive precisely in this perspective, in this reading. Input and output, technology and nature, raw materials and products, Man and community, with the economy becoming part of the whole again, part of human and natural society (Polanyi, 1944). The focus on ties and relationships allows integral ecology to be used to read the connection with social and institutional dynamics at all levels, as a powerful tool for analysing the resistance that opposes authentic care for the common home. What the neoclassical paradigm has lost sight of is precisely the complexity of links and interactions, which are instead at the centre of an ecosystemic view. This articulation will have to be embedded within the theoretical model, which must be inclusive and not exclusive. The path of the sustainable market economy, in fact, is not without strong weaknesses and contradictions. Firstly, because the hypothesised reduction of environmental impact will require radical economic and cultural changes, which are difficult to achieve in a capitalist market context. A possible reluctant attitude to economic change is intertwined with evidence that there are not enough resources for everyone to enjoy intensely competitive and consumerist lifestyles. Circular systems employ reuse, sharing, repair, refurbishment, regeneration and recycling to create a closed-loop system, minimising the use of resource inputs and the creation of waste, pollution and carbon emissions. Fundamental to this model is the conception of each waste as a source of use in another production process. Thus, the circular economy disrupts the normal logic of market production, re-evaluates the operating model, creating a flow of better and lower cost products, which diminishes economic divergence and competitive disadvantages. Diversity, sharing, versatility characterise the circular idea, which is therefore highly effective in today's scenario strongly characterised by an increase in demand for raw materials in the face of scarcity. Many of the resources essential for the economy are limited with the world's population continuing to grow, expected to increase by a further 2 billion by 2050. A circular economy and an integral approach will be needed. The 17 goals of the 2030 Agenda themselves are universally presented in a circular form to emphasise their interdependence and sequence.

The integral approach shows us the way to a comprehensive reading of these goals. Let us start with the first two: end poverty, end hunger, achieve food security, improve nutrition and promote sustainable agricolture. The demographic dynamic makes it very difficult to combat poverty: today, 9% of the world's population is considered to be in extreme poverty, which is defined as consumption of less than USD 1.90 a day at purchasing power parity in 2011. This is remarkable progress if we consider that in 1990 this share reached 35% of the world population, but the problem is that progress will be much slower from now on. An issue of great concern arises from the projected population increase, in concert with expected climate change, and changing eating habits (e.g. in favour of greater meat consumption). In emerging countries, global demand for food could increase by 70% in the coming decades. Increased demand will create further pressures. Finding solutions is very difficult as reducing food production is not easily feasible as any kind of reduction, either in production or distribution by any individual country or continent, would jeopardise food security in the world and contribute to rising food prices globally. Moreover, the richness of animal and plant species is threatened, especially in agriculture, by multinational corporations encouraging farmers to cultivate so-called 'high-yield' crops by planting monocultures, to the detriment of the hundreds of traditional varieties that are disappearing (Shiva, 2009). Here again, sustainability is not the answer. Interesting in this regard is the case of cities, of the world's metropolises. How to feed them in the future? Can they produce their own food? Urban agriculture systems can only provide a limited amount of necessary nutrition. But cities can obtain a large share of their food from the surrounding areas: according to the MacArthur Foundation (2017), 40% of the world's cultivated land, referred to as peri-urban areas, is located within 20 km of cities. A possible solution therefore lies in a radical change of the cultivation-supply-logistics model, with an integral approach of the circular economy. With goals 3, 4 and 5, we touch instead on three basic areas of civilisation, education, health and rights. Here the circular method finds perhaps its highest expression. Training, education, health, human rights and inclusion involve a holistic view, by definition. Service, common good, community spirit, connection to the past and preparation for the future. A decisive contribution can come from the world of associations, voluntary work, that part of civil society called the Third Sector. Third compared to State and Enterprise. It is the field of direct citizen participation in the care of the territory, of gratuitousness, of gift, with its ability to network, to build relationships to enhance social capital, to favour the multi-sectoral contribution of the various subjects. Politics will no longer be able to do without this contribution. The safeguarding of democracy will pass precisely through the extensive and articulated participation of categories as well as individual citizens in the care of the common home. With the following objectives, we move on to the major issues of water and energy, the availability and management of water, and to ensure that everyone has access to cheap, reliable and modern energy systems. The World Energy Council has defined the three dimensions of the energy transition: energy security, infrastructure reliability and energy equity. The first indicates the effective management of supply from domestic sources (e.g. solar panels) and external sources (e.g. hydropower plants); the second, the reliability of energy infrastructure and the ability of energy suppliers to meet current and future demand at all times and in full. Finally, energy equity, i.e. the accessibility of energy supply for all people and all peoples. It is not just a question of sustainability. We need a multi-year project, an innovative energy and industrial policy. Points 8 and 9 fit in perfectly with the previous one. First of all, stimulate economic, lasting, inclusive growth, full and productive employment and decent work for all. This is the theme of work, universal. Secondly, technological progress with its great opportunities and criticalities. Point 9 explicitly calls for a resilient infrastructure and fair, responsible industrialisation. In the age of digital does not only mean talking about technology, but about a cultural, social and economic dimension of human life. The concept of Industry 4.0, which concretises and implements the 'fourth industrial revolution', invades the perimeter of ethics, freedom, democracy. Artificial intelligence, 'machine learning', Internet of things. In the network of relations and connections between people that the Internet has initiated, and in which we now operate from any place and at any time with mobile devices such as smartphones or tablets, objects are being inserted. Work changes radically to the point of even being questioned (Marx and the reserve army, Keynes and technological unemployment, Schumpeter and innovation). The risk of inequality is realised in point 10. Internal and external inequality, between nations or areas of the world that have experienced in recent decades the economic paradox known as the 'resource curse'. Indeed, the Green and Sustainable Revolution has had no effect on the redistribution of economic power in Third World nations, and the poor have seen no progress in their purchasing power and land acquisition opportunities. Even a World Bank study stated that even a rapid increase in food production does not necessarily translate into food security. Warehouses may be full of rice and wheat but if the poor do not have the money to buy them, they will continue to starve. The Green Revolution brought undeniable benefits to those who already owned land, capital and political influence, while it accentuated marginalisation. Linked to this theme is that of the future of cities (Goal 11). There is talk of redesigning cities, the pride of the first two industrial revolutions and instead a huge problem of modernity, in terms of mobility, liveability and coexistence. Smart cities and city logistics, with attempts to reduce traffic, CO2 and noise emissions, waste management, improved quality of life, social bonding, increasing loneliness. With goal 12 we enter the heart of the market economic system, calling for the revision of production and consumption models. In recent years, socially responsible investment (SRI) has gained ground in business economics literature. SRI is a type of investment that takes into account environmental, social and corporate governance (ESG) criteria, which are useful for generating long-term financial returns. Here is green finance, green bonds with a green target, e.g. energy efficiency, waste treatment, eco-friendly construction. Or ethical finance or social finance that can become a tool for the common good (e.g. Solidarity bonds). On the consumption side, on the other hand, e-commerce has brought the consumer closer to the production and commercial network, while opening up the debate on conscious and informed consumption. Turning to Goals 13, 14 and 15, we deal with climate change and its consequences, the conservation of seas and oceans, the earth's ecosystem, forests, biological diversity. These are all related topics. Forests, for example, act on the climate, secure territories, are rain factories, self-regulating and self-healing. Their exploitation heavily affects the balance of the planet. The simple path of sustainability stops at the management of logging credits, at a more rational planning action. What is needed is an integral approach, which goes beyond mere reproductive or economic aspects, in the consideration of the forest as a natural community. The latter objectives are by definition far removed from any appeal to sustainability. Peace, inclusion, justice and world partnership call for the need for a project of coexistence, made up of cooperation, policies and philosophy. It is clear from all these objectives that there is a need to depart totally from previous economic thinking. In the traditional economic system, environment, territories, relationships were used and guided for exclusively productive, quantitative purposes. As we know today, this is no longer permitted. And making it sustainable is not enough. We must think of a new type of economy capable of containing all the different variables in a single inclusive, harmonious, universal vision. For an integral development, a deep ecology, a civil economy.

References

Bischoff, O. (2017). *"I concetti di transizione ecologica e di sviluppo sostenibile"*. https://etreserasmus.eu/?ConcepTIt/download&file=Transizione_IT.pdf

Bompan, E., & Brambilla, I. N. (2016). *Che cosa è l'economia circolare*. Edizioni Ambiente.

Mancuso, S. (2023). *Fitopolis*. La città vivente.

Meadows, D. H. (1972). *I limiti dello sviluppo*. Mondadori.

Polanyi, K. (1944). *The great transformation*. Farrar & Rinehart.

Shiva, V. (2009). *Semi del suicidio: i costi umani dell'ingegneria genetica in agricoltura*. Odradek.

Francesco Poggi Lecturer in History of Economic Thought, theories of business and organisations at the Department of Economics, University of Pisa. Lecturer UNESCO Chair in Economic System and Human Rights. Former board member of the IMT Doctoral School, School for Advanced Studies. Director of the journal "Nuova Economia e Storia", Member of the editorial board of the journal "Studi Economici e Sociali" and of the journal "Il pensiero economico italiano", Fabrizio Serra editore. A former member of the Board of Directors of Monte dei Paschi di Siena (2002–2011), he is the Creator and presenter of the Festival Economia e Spiritualità (since 2016) in Lucca, Florence, Prato, Pisa. Among the publications: *Economia e spiritualità*, (with contributions by S. Latouche, P. Barucci, S. Zamagni, L. Bruni, and others), 2018; *Europa vs Nazione. Economic integration, monetary union, international markets: the thinking of a century (1919–2019)*, Pisa University Press, 2019; *Valeur et utilité, comportement et choix. Ferdinando Galiani précurseur de Vilfredo Pareto*, in *Economie et politique chez Ferdinando Galiani*, Classiques Garnier (Paris), 2018; *Anthropological and religious dimension ecological transition and integral development: Economic Theory towards a new paradigm?* Economic alternatives (Sofia, Bulgaria), 2022; *From linear to circular model ESG criteria, EU policies and the new paradigm of civil economy*, Routledge, 2024; *Margherita Grassini Sarfatti, the economic thought of the jewish woman who popularize fascism*, Springer Natura, 2024; *Agenda 2030, ESG criteria and human rights: the way of the Civil Economy*, Palgrave Mcmillan, 2024.

Chapter 11
An Ethical Perspective on Linking Integrality, Sustainability, and Advances in Technology

Leon Miller

> *"To realize the true promise of the Earth Charter and the movement that supports it, we must reinvigorate the discussion of the ethics that created it and be much more explicit about the societal transformations that will be required to implement its underlying vision and principles than we have so far done" (Engles, 2014, xvi)*

Abstract Achieving the goals set out by the 2030 Sustainability Agenda involves more than strategies for making environmental protection an important aspect of development. Achieving the goals (including planning for the circular economy) requires an approach that is inclusive of human and social development, consideration of what is in the best interest of local stakeholders and their ecosystem, ethical considerations, addressing issues of corporate social responsibility, and consideration of moral responsibilities. Consequently, to realize the goal social action aims to achieve, the standard must include more than just a focus on wealth. Economic performance is one aspect of what is taken into consideration for improving the quality of social life. But, in addition, consideration must be given to the types of value that has worth in both tangible and intangible terms, what contributes to flourishing, what enriches and elevates the human experience, and what creates prosperity. The term prosperity is used to imply wealth but also because it has broader implications, which include thriving, doing and living well, and experiencing sustainable peace.

This chapter examines sustainability from an ethical perspective with a special emphasis on why the ethics of the sustainability concept play a special role in successfully implementing a holistic, balanced, inclusive, and multidimensional approach to future planning. This chapter argues that by including an ethical perspective in sustainability planning, it is possible to establish a standard for achieving

L. Miller (✉)
Tallinn University of Technology, Tallinn, Estonia

S. Patnaik et al. (eds.), *Circular Economy and Sustainable Value Creation through Eco-Innovation*, Circular Economy and Sustainability,
https://doi.org/10.1007/978-3-031-88516-7_11

multi-dimensional aims, which include increasing human and social well-being. In other words, if sustainability is a value that a society aims to achieve, it is possible to make predictions about the value orientation, ethics, and practices that are best for achieving its sustainability goals.

Keywords Foresight · Integrality · Value orientation · Operational standard · Value standard

11.1 Introduction

Achieving the goals set out by the 2030 Sustainability Agenda involves more than strategies for making environmental protection an important aspect of development. Achieving the goals (including planning for the circular economy) requires an approach that is inclusive of human and social development, consideration of what is in the best interest of local stakeholders and their ecosystem, ethical considerations, addressing issues of corporate social responsibility, and consideration of moral responsibilities. The prior approach to economic development proved to be inadequate, although planned by what at the time were regarded as the world's top development experts. This fact is especially significant considering that, in some cases, the top-down approach prescribed altering sustainable practices that had been in place for thousands of years—many of which were already based on a circular economy. That is to say that, in some instances, those practicing traditional sustainability methods were called on to modernize their traditions to bring them into accordance with the industrialization model of the more advanced societies. The prior economic development paradigm focused on generating wealth by means of production and consumption, increasing quantitative outputs, and a focus on the individual economic agent. However, they did not foresee that the extensive exploitation of natural resources to support mass production and the increase in consumption would create resource depletion, environmental devastation, contamination of water and soil, polluted air, and damage to our atmosphere.

We now realize that the prescriptions of the development experts only exacerbated environmental problems and climate challenges while they did little to improve social conditions and the quality of life. This prompted Nobel Prize Winner—who is now serving as the Chief Adviser of the interim government of Bangladesh, Muhammad Yunus—to stress that "Mainstream free-market theory postulates that you are contributing to society and the world in the best possible manner if you just concentrate on getting the most for yourself. This creates a very deep problem. Mainstream free-market theory suffers from a 'conceptualization failure', a failure to capture the essence of what it is to be human" (Yunus, 2008, 8). Thus, it is becoming increasingly apparent that a focus on defining development in terms of wealth and consumption not only hinders the ability to achieve the sustainability goals but, as well, the ability to provide a better quality of life in multiple dimensions (e.g., increasing life satisfaction and a higher level of social capital, which means less corruption and crime and thus higher levels of peacefulness and gross national happiness). Consequently, there is growing recognition that because the nature of the problem is systematic, resolving the problem requires an integrative approach to systems (i.e., an approach that draws from the relationality concept, is shaped by a System's Theory perspective of the interconnected web of existence, and one that takes a participatory approach when applied to a context).

Now, at the height of the sustainability and circular economy discourse, societies are once again called on to consider that their future growth and development might require incorporating the technological advancements of the developed countries (e.g., Industry 4.0, eco-technologies, artificial intelligence, and robotics). Because of this serious consideration is needed to determine how to plan for the best of all possible futures if planning will necessarily require blending nature (and eco-aesthetics) with maintaining the standard of living of advanced countries and advances in technology. Thus, the path we choose for our future is in serious need of rethinking. For, as Eistein is credited with saying, we can't solve our current problems by continuing to plan for the future and operate on the basis of the same way of thinking that created the problems we are now trying to overcome. "A new type of thinking is essential if humanity is to survive and move to higher levels" (Einstein 2007, 383). Therefore, there is a need for visionary foresight that shifts the emphasis from linear to circular and from an approach where economics dominates to an integrative approach that plans for growth and development in multiple dimensions.

What the members of a particular socio-cultural context consider to be growth (e.g., human, social, or sustainable) depends on how they define "the good life" and, as well, the ultimate value that they aim to achieve by means of their collective social activity. Following this line of thought, sustainable human and social development can be regarded as the manifestation of the collective value orientation that the culture establishes. This determines the normative basis for its sustainable flourishing and well-being. A collective value orientation, in this sense, is a society's understanding of what outcomes it should aim for in order to achieve its goals. In addition, adherence to societal norms ensures that the means the society undertakes to achieve its aims are consistent with its principles. That is to say that realizing value is the ultimate aim of social activity, and values determine the type of behavior undertaken as the instrumental means of experiencing that value. Sustainable social flourishing and well-being occurs when the value pursued and the instrumental means of achieving it result in the long-term maximization of the benefits the members of a social group are able to produce and enjoy in their relationship with each other and with the environment. In this respect, the idea of linking integrality and sustainable development involves normative claims regarding the most effective and efficient means of realizing social aims. Social aims are generally defined as social stability, cohesion, prosperity, flourishing, and sustainable peace.

A value orientation shapes the operational standard for social activity. The operational standard determines the performance measure chosen to evaluate the effectiveness of the instrumental means for achieving social goals. That is to say that a value standard establishes the framework for solidifying the connection between the instrumental path (or values chosen) and the good life a society aims to achieve. Consequently, to realize the goal social action aims to achieve, the standard has to be much more inclusive than just a focus on wealth. Economic performance is one aspect of what is taken into consideration for improving the quality of social life. But, in addition, the consideration must be given to value in connection what has worth in both tangible and intangible terms, what contributes to flourishing, what enriches and elevates the human experience, as well as prosperity. The term prosperity is used to imply wealth but also because it has broader implications, which include thriving, doing and living well, and fortune (i.e., fortune also includes the experience of good fortune and what it means to be fortunate). Therefore, as you might expect, by determining the value orientation for sustainability planning, it is possible to establish a standard for achieving multi-dimensional aims, which include increasing human and social well-being. In other words, if sustainability is a value that a society aims to achieve, it is possible to make predictions about the value orientation, ethics, and practices that are best in order to achieve its sustainability goals.

Social activity planned in this manner is sustainable in that the value pursued is likely to be both integrative and multidimensional, which means that it will promote human and social development, increase public value, improve quality of life, and increase prosperity. This requires understanding the type of value commitments that enable socio-cultural groups to experience value in terms of the triple bottom line (i.e., social psychological, economic, and environmental). It proposes that the members of a social group aim at maximizing *the good* in psychological, sociocultural, eco-aesthetic, and economic terms. Goodness is also a basic aim that social action attempts to realize because "All socio-political action has in itself a directedness towards knowledge of the good: of the good life, or the good society" (Strauss, 1957, 343). By making the goal of development the experience of goodness in multiple dimensions, planning necessary involves an ethical dimension in that goodness, in terms of its total scope, becomes a value outcome that social action ought to aim for. Thus, the integrative framework can be described as a holistic approach that has integrality, optimal performance, and goodness as value standards.

The integrality concept proposes an expanded view of development, which includes optimizing the performance of individuals, private organizations, and public institutions. That is to say that sustainability planning does not merely focus on increasing the market goods available for consumption. Planning on the basis of the integrality concept is a strategy for establishing complementary interactions between the members of society, maximizing the benefits they co-create and enjoy in social relations, and ensuring that their productions have a positive impact on the environment. The integrality concept emphasizes that because of social interconnectedness and interdependence, to achieve the optimal development and performance of social entities (e.g., individuals, social systems, and ecosystems), social interactions must be complementary, based on equal respect, and mutually beneficial. The idea of realizing optimal levels of performance is a desired good, which becomes a collective value preference that establishes a constitutive normative principle of society, and this realization should not be hampered because of a narrow view on development that overemphasizes production and consumption (Nussbaum 2000, 74). However, the idea of the right of entities to enjoy natural growth, flourishing, and well-being also applies to the ecosystem upon which society relies for sustainable growth.

This chapter argues that all social systems operate on the basis of a particular value orientation, and social action is based on certain values and principles. Therefore, improving the overall performance of a social system requires ensuring that it operates in accordance with values and principles that are effective for achieving the type of value that ensures the system sustained flourishing. As a matter of social planning, improving performance in ways that ensure sustainable flourishing and growth requires two ways of dealing with the issues of value, values, and evaluation. First is a descriptive analysis of the value that society aims to achieve by describing it in normative terms and, as well, explaining what justifies the reasoning for such normative claims. This provides normative guidance for future planning, establishes the basis for the *end value* desired, and enables planners to "Diagnose value conflicts, assess policies (actual and possible), and to validate or refute valuations placed on development performance" (Goulet 1997, 1168). Second is matching the value orientation with practical economic commitments to ensure that they reflect sustainability and circularity.

Therefore, this chapter examines sustainability from an ethical perspective with a special emphasis on why the ethics of the sustainability concept play a special role in successfully implementing a holistic, balanced, inclusive, and multidimensional approach to future planning. A key question addressed in this section is how the factors affecting sustainability compare to those that improve the quality of life. Thus, this chapter is concerned with determining the factors that would integrate achieving social aims with sustainability. Of course, advances in technology are an important contributor to economic growth. In particular, private organizations and public institutions are increasingly relying on artificial intelligence because of its effectiveness in improving efficiency. Therefore, significant consideration is given to the processes and practices that minimize harmful impacts on the environment by means of the circular economy (i.e., reducing waste, risks, and pollution and, as well, new methods for obtaining and using natural resources more effectively and efficiently, while, at the same time generating new and sustainable competitive advantages). The concluding section—which is entitled An Ethical Framework for Human, Social, and Sustainable Development—argues that by drawing from sustainability ethics it is possible to devise a conceptual model of what planning for the best of all possible futures should aim to achieve. Such a model emphasizes avoiding adverse effects that could result from inadequately balancing sustainability, human and social development, and improving the quality of life.

11.2 An Ethical Perspective on Linking Integrality and Sustainability

> *"Human development is primarily about being more, not having more. It affirms the values of reverence for the mystery of being, gratitude for the gift of life, and humility regarding the human place in nature"* (Earth Charter, 2000, 53).

Environmental awareness reached a new height in the 1960s. In part, this was due to the slogans of the counterculture movement. In addition to peace, love, and "go with the flow" the slogans of the counterculture movement included "back to nature". In fact, it was the rise in environmental consciousness in the 1960s that gave birth to a significant feature of the environmental movement, the first celebration of Earth Day in 1970. In addition, the social movements of the 1960s included feminists who argued that the dualistic split between the rich and the poor, minorities and the dominant members of society, and men and women, as examples, are all related to a dualistic split that is also evident in the way we relate to nature. Consequently, the feminist perspective on environmentalism (i.e., eco-justice) was aligned with the integrality concept in terms of promoting healing, the holistic development of individuals, and a more egalitarian social and world system, and they proposed a means by which humanity can be better integrated within the fabric of existence. Therefore, integral eco-feminism attempted to resolve the problem of fragmentation evident in many aspects of the human experience that result in the problem of dualism (nature-human, body-mind, fact-value, etc.). In short, integral eco-feminists sought to "Contribute to the philosophical base for new ecopolitics and eco-economics; to speak out clearly but without malice against those who further policies of injustice and ecological ignorance; to act with a pure mind and pure heart; to celebrate with gratitude the wonders of life on Earth; and to seek intimate communion with the natural world" (Spretnak, 1990, 14). It was in this respect that the feminist eco-justice movement championed ending all forms of oppression, exploitation, and dominance.

The emergence of eco-feminism was then accompanied by the early attempts at conceptualizing integral environmental ethics. Feminists proposed a shift from control, domination, and exploitation to a partnership ethic, which would maximize the good created for both humanity and the ecosystem that humanity depends on. This ethical perspective places a primacy on the ecological relational dynamics that occur between the parts of nature rather than on the primacy of any single entity in the interaction. Eco-centric ethics make each individual entity—whether male or female, white or black, the human or the nonhuman aspects of nature—equal in value to the other aspects of the overarching whole, with each deserving equal respect and rights. In other words, it places the ethical emphasis on increasing the human ability to care for others (i.e., caring for those who are alike and those who are different) and extending that same type of care to the environment; increasing the extent to which people care for ethical universals (e.g., justice, rights, and natural laws); and increasing the extent to which such practices of care are evident in interpersonal, organizational, institutional, and intercultural interactions (Merchant, 2005, 82).

"The eco-justice movement—initially characterized as merely being the outcry of an internationally coordinated group of nature lovers—gained support as a result of the increasing assertion, on the part of scientists from various disciplines, that the current approach to economic development has put the earth and its peoples on a non-survivable course" (Miller, 2020, 582). Consequently, environmentalism and climate change increasingly became scientized, politicized, and securitized. This growing awareness of broadener concerns regarding environmental problems and climate change established a linkage between "Four social science research programs: peace studies, security studies, development studies, and environment studies", all of which play a role in development planning, human security, and sustainable peace but have had limited interaction heretofore (Brauch & Spring, 2016, 3). Therefore, this more inclusive and integrated approach to development was regarded as essential for realizing the initial aim of the post-colonial movement, which was to create sustainable and peaceful social orders by liberating them from domination and exploitation.

The global environmental movement gathered momentum as a result of the force of The Earth Charter. The drafting of the Earth Charter was the outcome of a six-year worldwide process (1994–2000). The Earth Charter is the outcome of leading scientists, scholars and researchers, philosophers, social stakeholders, and leaders of the world's faith traditions collaborating to establish fundamental principles for building a just, sustainable, and peaceful global society in the twenty-first century. The declaration outlined the values, principles, and ethics that could shape the aims of a global environmental partnership for the wellbeing of the human family and the larger living world. The charter proposed four basic principles for shaping the future—all of which made reference to sustainability in addition to the established economic development paradigm: (1) respect and care for the biological community and for ecological integrity; (2) social justice and economic democracy; (3) non-violence; and (4) peace. The aim of the Earth Charter was to make environmentalism a shared value and sustainability a common goal.

The charter aimed to make clear the connection between troublesome global issues and environmentalism. In fact, what was apparent is that the charter articulated a vision of world leaders shifting their focus from issues that cause conflict between them to taking steps to resolve what the UN stated is "One of the greatest challenges of our time, the adverse impacts of which undermine the ability of all countries to achieve sustainable development" (UN, 2021, 7). The UN claimed that environmental and climate change problems are among the most important issues of our time not only because of the magnitude of the problems but because the problems are indeed a reflection of the tendency of attempts at dominating and controlling other aspects of existence. The tendency to attempt dominating and controlling other aspects of existence is based on the belief that "Certain "facts of life" are considered universal and ahistorical: that desires are endless and public resources scarce, people are fundamentally selfish and will fight each other for what they need, and that life is largely a zero-sum game characterized by competition where the victory of one is mostly won at the expense of another" (Escobar et al., 2024, 3).

The Earth Charter articulated a vision of a new global order in which the established substantialist view of existence would be replaced by a relational stance on ontology. Substantialists view discrete autonomous entities as the fundamental component of existence—meaning that the discrete entities are naturally shielded from being affected by their engagement with other entities. This perspective on the discrete autonomy of substances is based on the billiard ball model of the fundamental entities of existence. The relational view holds that things "Are not assumed to be independent existences present anterior to any relation, but they gain their whole being first in and with the relations that are predicated on them. Such 'things' are terms of relations, and as such, they can never be 'given' in isolation but only in community with each other" (Cassirer, 1953, 36). A relational view of ontology refers to the nature of existence by using such terms as interconnectedness and interdependence as well as terms that had been long established by the world's wisdom and faith traditions, such as the mutual identity and interpenetration of all phenomena, interdependent co-arising, and the interconnected web of existence. The relational turn in the social sciences, as an aspect of the movement toward a global sustainability partnership, aimed at grounding international law, intercultural relations, and development economics in the principles of natural law that promote harmony with the forces shaping the natural order.

The Earth Charter further clarified its integrated ethical agenda to resolve the critical challenges confronting humanity by introducing conceptualizations of relationality as a global partnership. The new global partnership focused on integrating the environmental, social, and economic dimensions of our global existence by proposing a framework for economic development that is not primarily oriented towards short-term economic gain but towards the flourishing of all of life. It also aimed to establish a regulatory framework for global relations, which contributed to a heightened interest in the relational turn in the social sciences (i.e., a universal ethical perspective on harmonious relations with others and nature that is also referred to as Cosmopolitan environmentalism). It highlighted the ethical significance of holistic well-being, integrality, and relationality. One of the main contributions of the charter is capturing of the character of present challenges, in that it "Recognizes that the goals of ecological protection, the eradication of poverty, equitable economic development, respect for human rights, democracy, and peace are interdependent and indivisible" (Earth Charter Initiative, 2000).

Consequently, there was a shift to a more inclusive perspective on development. The endeavor to shift to a more inclusive approach to development was supported by the release of Our Common Future by the World Commission on Environment and Development in 1987. Initially, this represented an effort to establish a global partnership for integrating environmentalism and development strategies. Thus, initially the transition to a more inclusive perspective focused on environmentally friendly ways of planning and engaging in development. The benefits resulting from improved economic performance would gradually trickle down in order to improve the conditions of all segments of society—a strategy that had inadequacies and ethical defects that would become increasingly more evident. This view was endorsed by Neoliberals, who envisioned ushering in a new, more technologically advanced stage of production (i.e., Industry 4.0). The fourth stage of the industrial revolution meant planning for consumption in ways that could be sustained, are environmentally friendly, that would reduce waste or more effectively recycle waste, and in ways that require new energy systems based on clean technologies that reduce the human impact on the environment while, at the same time, allowing for the continuity of economic development. This approach found favor with industrialists and international corporations because it demanded more efficient approaches to production, which would eventually lead to cost reduction. This established the dominant strategy for continued economic growth in much of the developed world, which subsequently shaped the ongoing discourse about development in underdeveloped and developing countries.

Thus, many experts and practitioners were convinced that the application of eco-innovations and new technologies are essential aspects of what development planning should aim to achieve. There was a community of experts attempting to find the best strategy for future planning and sustained economic growth who were convinced that it is not possible to provide the improved quality of life that so many of the world's 8 billion people desire without making use of the opportunities provided by the material production, consumption, and product lifecycle model proposed by Industry 4.0. The new technologies included artificial intelligence, grid technologies, eco-technologies, and integrated technological systems. For others, advanced technology is only one possible means of achieving outcomes that will improve human and social conditions in multiple dimensions. Therefore, the decision over whether or not advanced technology offers the preferred path is a matter of value preferences and the reasons for having such preferences, which makes it necessary to clarify the outcome that planning for sustainability is meant to achieve and, as well, the best means of achieving the desired outcome. This is especially true in reference to determining which means are best or preferrable for particular contexts. In other words, "A sustainable development path should be defined simply as one that can be sustained over some specified time period, and whether or not it ought to be followed is another matter. It should be treated, in other words, as a purely technical concept. One obvious rival injunction would be to seek the highest welfare for society over some specified time period" (Beckerman, 1994, 193).

The uncertainty regarding technology is due to the fact that "Technology provides its own twin-horned dilemma or paradox, being both a significant cause of problems and a potential source of solutions. Consequently, "Any effort to analyze technology for the suitability of its deployment inevitably encounters ethical dilemmas, many of them linked to sustainability concerns" (Kibert et al., 2012, 21). Ethicists, social theorists, and environmental experts, who are critical of the focus on advanced technologies, admit that "The 'technological imperative' is indeed unquestionable in its anthropological primacy and thus is integral to the human condition. However, modern technology has introduced actions of such novel scale, objects, and consequences that the enormity of its power forces upon ethics a new dimension of responsibility never dreamed of before" (Jonas, 1984, 6 & 203).

Resolving the ethical dilemma can be achieved by planning on the basis of a sustainability science in which the specialist applies science and engineering to move from our current problems to sustainable solutions by utilizing technologically advanced problem-solving processes. It is only by means of a framework for critically scrutinizing assumptions about knowledge generation and application, the impact of advances in technology on the human condition and the environment, and the extent to which advances in technology contribute to realizing "the triple bottom line" that we can move ahead to maximize the good outcomes they can produce and minimize their ill effects.

"Because there is a lot at stake from the new technologies, but uncertainties about the nature of the harms that could take decades to be resolved if they can be resolved at all, ethical reasoning is deeply challenged" (Brown, 2014, 51). Ethical reasoning regarding the use of new technologies involves determining their effectiveness in enabling a society to achieve its aims as well as the UN's Sustainability Agenda. This primarily involves a careful analysis of the interaction between humans, technology (i.e., virtual realities or augmented reality), and nature. The ultimate goal of an ethical analysis is to determine what actions are right or wrong in terms of contributing to integrality, human and social well-being, improving the quality of life, and, in short, the human good. "Emerging technologies offer the promise of higher productivity, increased efficiency, and decreased pollution, but many bring risks of new toxic chemicals and wastes and of major accidents of a type and scale beyond present coping mechanisms" (Our Common Future, 1987, 16). Thus, the commission that drafted Our Common Future stressed that an ethical analysis is essential to determining what actions contribute to human well-being and survival, successful development strategies, and the quality of life—indeed, to life on earth itself.

Consequently, there was a growing interest in an approach to development research that would produce the types of eco-innovations "That mimic nature or that rely on natural processes" (Kibert et al., 2012, 35). This gave rise to the technological-material focused sustainability model that is referred to as the circular economy. The circular economy is an approach to sustainable development that envisions a balanced blend of people, nature, and technology, which is based on the circular economy principle of being sure that products have no harmful effects on the environment throughout their entire lifespan—"from the cradle to the grave". The earliest proposition for the circle economy called for "All outputs from consumption constantly being recycled to become inputs for production" (Boulding, 2011, 7). The circular economy, in contrast to the onward and upward approach to the prior linear economic development paradigm, increasingly attracts the attention of governments, researchers, practitioners, and stakeholders as a necessary step toward achieving development goals.

The circular economy creates closed-loop systems in which materials are reused and recycled; waste is reduced; systems operate on the basis of biomimetic processes; and the circular economy employs strategies for increasing eco-efficiency. Biomimetics, or biomimicry, imitates nature's forms, processes, and functions by creating new eco-innovations and materials that solve human problems. Thus, specialists and practitioners anticipate, with much hope, the possibility of futuristic industrial practices that operate on the basis of what is taken from nature is returned in ways that are completely compatible and harmonious with natural patterns and cycles. The eco-innovations and technologies provided by the circular economy are important to developed countries because they contribute to achieving development goals in ways that also allow them to maintain their standard of living. The aim is to apply eco-technologies in ways that create an Ecotopia. Ecotopia is a vision of a society where people live well, enjoy prosperity, relate well to each other, experience social justice, and have a fair distribution of social goods.

Achieving this aim requires people and societies cooperating in a superordinate cause to realize the vision of a sustainable future rather than continuing to devote production toward destructive forces and practices by devoting billions toward creating more destructive forces that threaten the sustainability of human existence. By cooperating to realize the vision of a sustainable future, global stakeholders could avoid going further into global divisiveness, peril, destabilization, and an escalating tendency for things to fall apart. The issue is, what, exactly, is falling apart and what has become destabilized? Destabilization is the result of a center that continues to show evidence of domination and control by emphasizing gaining and maintaining the competitive or comparative advantage. And we, to date, have been inadequate in "Staking out stable ground around which new futures can be negotiated" (Zavestoski & Weigert, 2018, 26)? Staking out a stable ground upon which a secure future can be established is the ethical challenge of our time. From an ethical perspective, staking out a future that adequately incorporates integrality and sustainability requires addressing the very root of what is causing "a shaking of the foundations".

The endeavor to resolve our current problems by combining economic growth and environmentalism (i.e., coupling consumption and production with defining development in terms of a market-led green economy that will overcome the dichotomy between production and ecology), does not address the core of the problem. The core of the problem lies in a value perspective that prompts creating unsustainable socio-technical systems. Thus, overcoming the problem requires both a social and value transformation. "Transformation refers to a change from an existing state to one that is more desirable" (Miller, 2022, 172). Because a transformation involves progressive social activity, it is also referred to as a value transformation. Value transformation is a higher-order concept that involves inter-relational processes that enrich the eco-aesthetic dimensions of socio-cultural life, increase ontological security, improve human and social well-being, and elevate the human experience. Conceptualizations of value transformation stress that the combination of people, the economy, and the artifacts a system produces (including technology and society's structural components) are all embedded within a larger ecosystem. It is the interactions between all the components that shape the nature of the ecosystem. Transformational value results from generating knowledge of how to establish the optimal performance of all the contributing components (e.g., the individual members of society; the organizational, institutional, and structural components of society; and the ecosystem upon which they all depend).

Therefore, there was increasing recognition that the established paradigm was falling short of improving the human condition because it focused on increasing the material aspects of existence but did not pay adequate attention to human, social, and sustainable development. In other words, it focused on the cosmetic, apparent, or outer dimensions of growth and on quantitative outputs that could be empirically measured. Therefore, it overlooked enriching the aspects of the human experience that are less apparent but even more important (e.g., what enriches the inner nature of individuals, what increases human security, and the qualitative aspects of human and social existence). Achieving the optimal performance of the overall system (i.e., a system transformation) requires an integrative approach to development. According to the UN Declaration on Social Progress and Development, the purpose of government, political activities, and the use of political power is to "Ensure the social progress and well-being of the people, plan social development measurements as a part of its comprehensive development plans, and encourage and coordinate or integrate all national efforts toward this end, which might require social transformations" (UN, 2004, 80). Consequently, the economic development paradigm was superseded by the sustainability discourse.

The sustainability concept calls for incorporating the principles, values, and ethics of The Earth Charter, Our Common Heritage, eco-feminism, those of eco-centric societies and cultures, and the circular economy (Callenbach, 1975, 11–12). Conceptualizations of sustainability address and attempt to resolve the dichotomy between continued economic growth and the ecological reality of limits. Although technology is a feature of a sustainable approach to development, its use, practices, and operations are shaped by the ethics, values, and principles that are the basis of the Earth Charter, Our Common Heritage, and the UN's Sustainability Agenda. In short, sustainability calls for a vision of the future that is grounded in human-centered values, principles, and ethics, which indeed are, within themselves, sustainable. Realizing this vision requires a more equitable approach to resources integration and to human, social, and sustainable development. The point is that concentrating on more environmentally friendly approaches to produce and provide consumption choices within itself will not address issues related to human and social development, social justice, and human rights (Merli et al., 2018, 703–718).

A move toward a value transformation framework that was a step toward a human-centered indicator for sustainable development caught worldwide attention when Bhutan announced the importance of going beyond measuring growth merely in terms of material values. Bhutan proposed a plan for development that is in line with the integrality concept by asserting that happiness—as a measure of the outcome of development—is more important than GDP. Therefore, Bhutan introduced an approach to future planning that balances material pursuits with enriching the inner person, along with a value perspective that emphasizes living in harmony with others and nature. This approach to future planning was based on a cultural worldview in which the material aspects of society are a part of a pattern of relationships where each component plays a part in the well-being of the whole. "Since 2008, the country has used a metrics-based system in support of gross national happiness implementation, with conservation of the environment as one of the nine domains of gross national happiness" (World Bank, 2022, 2). In terms of a normative framework that society should aim to achieve, Bhutan established four essential goals: good governance, sustainable development, preservation and promotion of culture, and environmental conservation. However, the framework also emphasizes that social action is based on certain collective values, norms, and ethics, which should be taken into consideration when undertaking development planning.

But an equally, if not more impactful, framework for development was influenced by a strategy for simultaneously enhancing a society's ability to prosper, develop the capabilities of the individual members of society, improve the overall quality of social life, and implement effective and efficient sustainability measures (Nussbaum & Sen, 1993, 1). From this perspective, the measure of human, social, and sustainable development is providing individuals the ability to enjoy a good quality of life over a prolonged period of time, access to the type of knowledge that would contribute to their self-cultivation, ensuring that individuals have fair access to social goods so that they are able to live a satisfactory and socially respected lifestyle, and integrating quality of life with sustainability planning. The proponents of this capability approach complained that the Neo-liberal strategies that prevailed during the prior periods of the economic development paradigm viewed human well-being, social development, and sustainability as desirable biproducts of economic growth. However, the capability approach emphasizes that "Development is fundamentally an empowering process" (Sen, 249). From the capability perspective, the aim of sustainability is to increase the effective freedom of individuals by developing their capabilities and empowering them to live in accordance with what they value, which brings the constructive agency of people directly within the domain of environment-friendly developmental achievements.

From an ethical standpoint, an analysis of the economic development paradigm and its transition to sustainability provides the basis for developing a framework for what social action should aim to achieve in order to secure human and environmental flourishing. When analyzing the progression from environmentalism to eco-justice, from economic development to the circular economy, and now to an approach that focuses on integrating human, social, and sustainable development, it is clear that there are firmly established foundational values, principles, and ethics that can shape future security, peace, and flourishing. There are two means to determine how to achieve outcomes that would contribute to realizing both social aims and the goals of the sustainability agenda: an ethical approach that foregrounds a normative framework for the value outcome social action should aim for and an instrumental perspective that emphasizes the operative values for achieving the desired outcome. By applying these two means, it is possible to establish a framework that will enable a society to be effective in human, social, and sustainable development, or, in other words, to implement the types of social activity that are best for improving the quality of social life in multiple dimensions. The goal is to achieve the optimal performance of individuals (i.e., the development of individual capabilities), of private organizations and public institutions, and of the larger ecosystem that a society depends on for sustained flourishing. The point is that an operational standard should generally be used to help a society achieve what it really values most, to guide decision-makers in setting goals and making sure they are on target for achieving their desired outcome.

11.3 An Ethical Framework for Human, Social, and Sustainable Development

Linking integrality with sustainability is "Inevitably of far-reaching importance to the quality of life on earth—indeed, to life itself. Human survival and well-being could depend on success in elevating this to a global ethic" (Our Common Future, 1987, 308).

*In nature, change doesn't happen from a top-down, strategic approach. There is never a boss in a living system. Change happens from within, from many local interactions occurring simultaneously (*Wheatley, 2001, 2).

"Sustainability is often thought of as a long-term goal (i.e., a more sustainable world), while sustainable development refers to the many processes and pathways to achieve it" (UNESCO, 2012, 2). The member states of the United Nations all agree that achieving the goal requires collaborating as a global community to create planetary healing and future security. The world leaders agree to five fundamental principles that will contribute to bringing this about: (1) universality (it applies to all countries); (2) leaving no one behind (it considers all people irrespective of status and location); (3) interconnectedness and indivisibility (it needs to be implemented as a whole); (4) inclusiveness (every contribution is valid); and (5) multi-stakeholder partnerships to support implementation. The agenda contains an implementation framework of 17 sustainable development goals, 169 targets, and 244 unique social-ecological indicators (93 of which are environment-related). In other words, the goals, targets, and indicators are all parts of a framework for determining what to aim for; the measures for integrating and assessing different areas of human, social, and environmental development; and the key performance indicators with respect to certain human, social, and environmental concerns. However, unfortunately, as *The Sustainable Development Goals Report 2024* announces, we must face the sobering fact that less than one fifth of targets are on track, "Nearly half are showing minimal or moderate progress, and progress on over one third has stalled or even regressed. The report reveals that progress has ground to a halt or been reversed across multiple fronts, despite reaffirmed pledges" (UN, 2024, 2 & 3).

"While there is a large amount of debate on what needs to be done, the question rarely asked is: Why has hardly any effective progress been made" (Konold & Schwietring, 2021, 131). In other words, there is a need to address the discrepancy between the expressed commitment to the goals of the sustainability agenda and the aims a society is actually trying to achieve. Are societal goals and aims compatible with those of the sustainability agenda? Do the goals of the sustainability agenda contribute to realizing societal aims? And is it possible to determine and overcome what is hindering achieving the aims of both? Indeed, what seems to be most needed is an analysis of why societies are so intransigent in attempts to achieve sustainable human and social well-being, security, flourishing, and peace. Social theorists provide clarity on this dilemma by pointing out that "If sustainability refers to a substantial reorientation—and not merely a little less consumption, a little more efficiency in the use of energy and material resources, and slightly cleaner production—the task ahead requires no less than a profound change in social patterns that have developed over centuries" (Konold & Schwietring, 2021, 133). Although the environmental movement has, over the last 50 plus years, enjoyed considerable media attention and the approval of world leaders, international NGOs, and scientists, it faces the resistance of convictions deeply rooted in views on political economy, economic realism, and political realism that have dominated for several hundred years. In addition, this dominant ensemble of beliefs, socio-cultural norms, attitudes, and practices is anchored in a mindset that was established at the emergence of civilization and prevails even right up to today.

A historical analysis of social activity reveals that public authorities and the powerful elite typically share the conviction that social stability and growth require maximizing the utility benefits they can gain from external resources. Utility benefits are gained by accurately calculating how to achieve growth in terms of economic and technological power while, at the same time, ensuring the stability of social existence. Therefore, public authorities share the conviction that their strength, prosperity, and security require the type of growth provided by economic development strategies, and societies view the relationship between what is internal and what is external from the perspective of self-interest. Historically, the environment is viewed as an external resource for strengthening the economy. Whatever problems occur in relation to factors that are external to the system can be solved by further economic and technological development. The greater the internal complexity of a social system, the more it tends to define itself and its processes as distinct from the outer things that are not a part of its inner existence and by creating a protective boundary between itself and the other things in its outer circle.

The mindset that dominated perceptions of internal and external relations at the earliest stages of social existence is characterized as undifferentiated. The members of a social group shared a strong close bond, and their sense of relationship extended to the environment (and, in many cases, even their sense of kinship extended to the environment). When the stage of roaming freely without boundaries was transformed during the village stage of existence and social groups became territorial, there was increased awareness of a need to protect the social group from external threats. This was followed by kingdoms and empires, all of which devoted a tremendous amount of resources to building barriers of protection and manpower toward establishing elaborate security systems to both protect themselves against outside threats and to ensure being able to acquire what was needed from the environment. Consequently, even up to today, social agents tend to create a differentiation (i.e., a binary self-other distinction) between the inner system and the external resources needed satisfying needs. "More exactly, the theme of sociological investigation is the difference between the social system and its environment. In other words, the theme is the world as seen through the system of society, i.e., with the help of distinctions by which the system of society differentiates itself from the environment" (Niklas, 1989, 7).

In a similar fashion, the contemporary perspective on development is embedded in a conviction, rooted in both Modernity and the industrial revolution, that links the use of science and technology to acquire needed natural resources with what advances socio-economic conditions. In addition, economic development is equated with notions of power, wealth, and freedom (Bacon 2011, 29 & 253; also see Bacon 2011, 7–8). The intellectual and power elite, along with the political and economic experts of the major zones of civilization, envisioned that science, technology, industrialization, mass production, and consumption would progress humanity past its feudal history into an era when governments represent the will of the people and there will be unending material progress, prosperity, and growth. This laid the foundation for *the laws governing development*, whose primary corpus was *The Nature and Causes of the Wealth of Nations*. The conviction was that a moral or ethical understanding of progress necessarily draws from a scientific and empirical analysis of what contributes to growth, what creates prosperity, and what truly has worth and the best approach to development. It was believed that science, mechanical arts, and technology would ultimately progress humanity toward better living standards (enjoyed by a larger number of people), independence, and sustainable peace. However, the focus was on complying with the epistemological, ontological, ethical, and axiological guidelines laid out by economic experts, empiricists, development experts, and technocrats. This way of thinking was rooted in convictions that date back to the emergence of civilization, when political authorities were convinced that achieving the social goals of stability, security, and survival required increasing economic, production, technological capabilities, military power, and its comparative advantage.

Historically speaking, the literature on system differentiation emphasized that the power, security, and stability of a social system lie in its ability to grow in complexity, which results in creating greater distinctions. This means that "The conceptual model of a whole that consists of parts and relations between parts was superseded by a model that focuses on establishing selective boundaries. Action in such a society relies on a complex network of selective boundaries to reduce contingencies" (Niklas, 1977, 30 & 31). Things that are external represent contingencies that are best managed with the power of predictability and control. Convictions regarding power and control (the power to predict and control both the actions of individual social agents and the environment) are based on the belief that the likelihood of surviving whatever threats society faces from external forces is enhanced by increased predictability. The forms and processes of differentiation involve specifying what aspects of the external environment have utility value for the internal system, which has three consequences: (1) nature is seen as juxtaposed to the human experience; (2) nature is viewed for its utility value thus as a resource for economic and technological development, which also tends toward anthropocentrism; and (3) a social subsystem (for example, the economy) can play a selective role in directing the development of the overall ecosystem. "Any attempt to restructure the dynamics of social systems, for example, towards less resource use, less pollution, or less consumption, will be answered by the different social subsystems in the only way each system can operate. The economic system, for example, will turn it into a question of paying/not paying" (Konold & Schwietring, 2021, 135).

As a result of such strongly held convictions, economic development and industrial production continued to be regarded as the primary focus—an ethic—for all social activity. Such an ethic concentrated on benefits for individual social agents, which was presumed would ultimately create benefits for the overall society. In short, this meant that increasing economic, industrial, and technological power established the "Doctrine, and related social practices of international development. The doctrine was that all, or virtually all, economic and social problems have an economic solution" (Howard & King, 2004, 40). Consequently, economics extended its role from being a component of a social system to being regarded as the driving force of social and economic development. Those sharing this conviction claimed that development aims could be achieved by extending the reach of economics into more spheres of social life and by maximizing the reach of liberal market mechanisms. This was coupled with the conviction that social activity should be subject to economic theory and methodology, plus that the social good could be maximized by improving GDP.

This meant that the standard for development turned out to be the practices of the technologically advanced countries of the developed world. Thus, societies continued to operate under the assumption that higher levels of economic production and consumption equate to higher levels of social stability, security, and power. Social agents who hold such convictions, while on the one hand agreeing to the need for environmental protection, on the other hand persist in development strategies that enable them to maintain their economic, military, and technological power. Consequently, prioritizing comparative and/or competitive advantages makes it nearly impossible to resolve the problem of inequalities. The problem lies in a subsystem's claim that it is the "queen of the social sciences". This results in a discrepancy between its exclusive epistemology and methodology and the inclusiveness required for social science research. Epistemological and methodological exclusiveness accentuated the persistence of processes and forms of differentiation (i.e., the typical functional logic of social systems) and, as well, placed the emphasis on autonomous individuality. This perspective proves to be inadequate for social science analysis because it claims that "All social phenomena resolve themselves into decisions and actions of individuals that need not or cannot be further analyzed. A view which, of course, is untenable so far as it implies a theory of the social process. It fails to take into account the formative influences of environments, group attitudes, and group valuations. This is one of the reasons why the 'ethical' aspects should be emphasized" (Schumpeter, 2006, 885).

By narrowing notions of worth to economic value, the prior paradigm actually overlooked the broader perspective on value that is embraced by social value theory. Indeed, economic value theory established an exclusive perspective that is regarded to have epistemological and methodological supremacy, thus resulting in an incongruence between social value theory and economic value theory. Consequently, the prior development paradigm contrasted with *methodological holism.* The need for a more inclusive approach, which was demanded by the sustainability discourse, prompted a greater consideration of methodological holism. Holism, as prescribed in the sustainability discourse, promotes a Systems Theory perspective on both nature-human and self-other relations. Holism is the theory that "Social entities are unique "wholes" that cannot be reduced to the actions, beliefs, and values of individuals. Holism is the doctrine that a social whole is more than the sum of its individual parts—or, alternatively, that the whole is in some way prior to the individuals who comprise it" (Smith, 2013, 200). This means that there are at least some aspects of social phenomena that are not reducible to or adequately explained in terms of individual activity (see Fig. 11.1 below). The proponents of holism propose that it enables a "Recovery of the truth of who we and the rest of life most essentially are and, as such, a timely prophetic declaration of the universal ethical laws in terms of which we ought to pattern our lives" (Engles, 2014, xvii). Far from proposing an ideal utopia, this ethical grounding is based on the very principles of the underlying natural order, which establishes natural law, and that stands in judgment of every finite and limited human understanding and practice.

Overcoming the challenge of the systemic influence of political and economic realism in "Decision-making, in the effort to implement the concept and principles of sustainable development, requires developing an integrative approach that bridges the multiple dimensions of development, both institutionally (as policy silos) and methodologically (as disciplinary paradigms)" (Dalal-Clayton & Sadler, 2004, 3). This means developing a theoretical and methodological framework for overcoming the obstacles by blending positivist and normative strategies for achieving

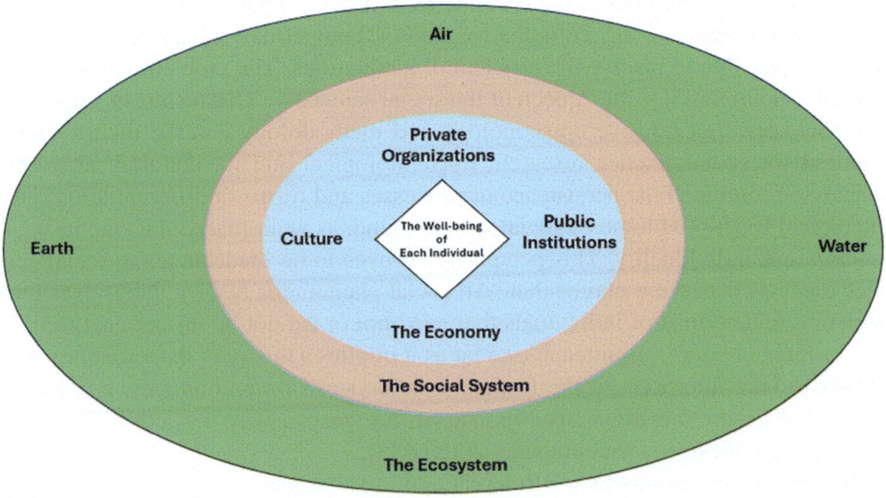

Fig. 11.1 A social system is regarded as existing within an ecosystem. Or, to put it in other words, a social system is encompassed by an ecosystem. In addition, each subsystem is one component of a social system. Although this model recognizes differentiation, it also points out that in flourishing systems the resources of each component are integrated to increase the efficiency and effectiveness of each subsystem, which creates sustainable benefits for the entire social system and its ecosystem

development in multiple dimensions. This includes developing a theoretical and methodological framework for overcoming the deep-rooted obstacle of a social subsystem (for example, the economy) playing a selective role in directing the development of the overall system. The ethical principles of the integrative approach prescribe a means by which future social action minimizes the dichotomy between realist political and economic values and the ethical demands and value orientation of a multidimensional and holistic approach to future planning. This makes the integrative approach desirable because it is a strategy for achieving desired social outcomes based on ethical principles while, at the same time, providing a value orientation that enable society to realize its aims.

Indicators of development are often focused on improving the relationship between the economy and the larger ecosystem in which it is a component. However, the economy itself is only one component of a larger social structure. Each component of a social system must support and reinforce the others if the system is to experience growth in multiple dimensions. In other words, "What we measure shapes what we collectively strive to pursue—and what we pursue determines what we measure, which impacts the way in which our societies look at themselves and, therefore, on the way in which policies are designed, implemented, and assessed" (Stiglitz et al., 2009, 9). The point is that a social system can be experiencing growth in one of its dimensions. However, that growth in one dimension does not contribute to growth in other dimensions, thus does not support the sustainable development of the whole system. Attempts to develop a more comprehensive approach to positive social transformation resulted in the construction of frameworks like the genuine progress indicator, the sustainable social development model, the human development approach, the global multidimensional poverty index, and the measure of life satisfaction. Such models aim to provide an approach to transformational growth that generates social and economic well-being, health, safety, security, happiness, and sustainability. Countries prize such frameworks because it helps them to determine how best they can advance the pursuit of their goals through collective actions.

It might still be the case that some countries will prefer focusing on a single dimension of growth, believing that it is more advantageous in the long run. Such countries will continue to prefer measuring growth in terms of economic strength, which they believe is driven by industrial and technological development. The assumption is that societies with a higher level of GDP are better off. This preference will be accompanied by the conviction that throughout the history of civilization, humanity has always faced development challenges. In all instances, throughout history, "progress" has always been the best means of managing those challenges. Progress is always driven by economic development and advances in technology, and a quantifiable measure of progress is the rate of GDP growth. However, all the indicators reveal that there will be crucial social drawbacks as a result of a failure to value what counts the most for wellbeing in multiple dimensions. For example, development focused on a single dimension might very well improve economic conditions and promote technological development but fail to realize the total scope of what social action aims to achieve (OECD, 2011, 14–18). For example, the United States has had the best GDP for an astonishingly large number of years.

However, its scores on levels of anger, stress, and violence have steadily increased for five years in a row while, at the same time, scores on optimism, level of happiness, and peacefulness have steadily decreased for each of those years. Thus, people are wealthier in terms of per capita GDP but less satisfied with their quality of life, and they report that the focus on the pursuit of wealth is not making them happier, which is indeed an ethical issue.

"No sustainable improvement in human well-being is possible without growth. But it is also wrong to suggest that high economic growth rates will automatically translate into a higher level of human development" (ul Haq, 2003, 9). Individual social agents may profit from economic development to such an extent that GDP rises enormously without reducing poverty, increasing human development, or improving quality of life. This is why most countries prefer improving their social conditions in as many dimensions as possible. Theorists recognize that to integrate the multiple dimensions involved in improving social and economic well-being, there needs to be overarching strategic performance measurement tools. Consequently, sustainability theorists realized that there is a need for an integrative and multi-dimensional approach to take care of all the ethical dimensions of equity. This includes an ethical analysis of the international aspects of the interdependence of resource generation, use, and fair distribution. This has prompted the introduction of an ethical framework for sustainable development that integrates a society's endeavor to achieve its aims in ways compatible with the sustainability agenda. The ethical balance framework builds on the Balance Scoreboard model established in private industries.

Countries that aim for strategic success in multidimensions prefer integrating the resources of society (e.g., human, socio-cultural, economic, technological, and environmental) by means of a social value creation network to generate benefits for each component of the system. Consequently, sustainability theorists developed a new tool that "Combines the resource integration of networks with multi-objective optimization to develop sustainable clusters. The model allows the integration of all resources in a cluster while simultaneously optimizing the economic, environmental, and material conservation performances" (Ahmed et al., 2021, 11). Resource integration contributes to multi-objective optimization in that it engages stakeholders in a value creation network that improves social and economic conditions in ways that help to meet sustainability objectives. Resource integration, as a multidimensional approach to improving social conditions, provides growth in terms of what not only increases value in economic terms but also increases public value (i.e., makes life as a whole better).

Making life as a whole better involves consideration of the fundamental principles for human and social development, which are inclusive of the role of moral development in the design of "the good society". The fundamental principles assert that, from an ethical perspective, a comprehensive approach to development goes beyond providing for the material needs of individuals to include increasing the likelihood that individuals can experience goodness, an improvement in the quality of their lives, and an increase in life satisfaction (Ibn Ashur, 2006, 91). Of course, on the one hand, an ethical stance on development is significant because one of the primary aims of society is to minimize—or even completely eliminate—the conditions that threaten to diminish the quality of life. However, on the other hand, minimizing the conditions that threaten to diminish the quality of life requires the concerted, collaborative cooperation of individuals, private organizations, and public institutions, NGOs, and intergovernmental organizations. Therefore, there are multi-level relational dynamics that, when viewed from an ethical perspective, involve a moral responsibility to engage in a multidimensional approach to sustainability planning.

Resource integration allows for a more holistic, dynamic, inclusive, and participatory approach to creating future social reality. Because of its effectiveness in creating social and economic benefits in multiple dimensions, it is often referred to as creating public value (Moore, 1995), the co-creation of value (Prahalad & Ramaswamy, 2004), and creating shared value (Porter & Kramer, 2011), and now even creating sustainable value (Hart & Milstein, 2003). Applying the value creation concept to knowledge-generated social actions necessitates engaging in collaborative, multidimensional, constructivist-type interaction processes to maximize the benefits for individuals, private organizations, public institutions, and society as a whole (Vargo & Akaka, 2012, 207–217). Performance indicators reveal that there is a clear correlation between effective resource integration and scoring high on the Quality of Life Index, ranking high in human development, success in achieving the sustainability goals outlined by the 2030 Sustainability Agenda, and above-average scores on GDP. Therefore, the performance measures indicate that a broader perspective on sustainability contributes to achieving social aims in multidimensions.

Resource integration is so effective in improving performance in multiple dimensions because it is a means of developing generative power. "That is to say that an analysis of networks reveals that their significance lies in the fact that they generate future social and economic value. In this respect, they also play an essential role in social power, social cohesion, the power of special interest groups, and the power of open participatory integrated networks" (Miller, 2022, 86). Therefore, there is widespread acknowledgement that resource integration networks could ultimately overcome some of the structure-agent relations inadequacies of the prior approach to development. Resource integration networks are linked by collective, dynamic processes that involve the members offering and exchanging various types of essential resources to the participants in the network. The actors are defined as providers (i.e., resources applied for benefit of all the components of the system). The network serves the purpose of enhancing the performance and wellbeing of individual social agents as a partial function of improving the wellbeing of the entire system and its ecosystem (Vargo & Lusch, 2017, 49). In this respect, the normative function of a social system is to connect people, resources, technology, and information through a value creation network with the aim of co-creating value for all the components of the system.

Table 11.1 below indicates the effectiveness of implementing the integrative approach to sustainability. The factors that make this possible, thus factors that can be used in other social contexts to achieve similar results, can be summarized as follows: (1) achieving a high level of individual well-being and higher levels of

Table 11.1 The analysis indicates that multidimensional success is the result of implementing value creating social networks that establish measurable objectives. The effectiveness of the value creation network is based on their ability to raise awareness, increase human well-being and the political accountability that makes this possible, establish effective performance metrics, provide feedback, and increase stakeholder involvement

Comparative Scores on various dimensions of development (most scores are from 2020 but also from 2024 where indicated)				
Country	Quality of Life	Human Development	Achieving the UN's 2030 sustainability goals	The World Happiness Report
Finland	Scored # 1	Scored # 12 with a score of .942	Scored # 1 (2024 report)	Scored # 1
Switzerland	Scored # 2	Scored # 1 with a score of .967	Scored # 24 (2024 report)	Scored # 3
United States	Scored # 14	Scored # 20 with a score of .927	Scored # 26 (2024 report	Scored # 18
Japan	Scored # 16	Scored # 24 with a score of .920	Scored # 18 (2024 report	Scored # 51 (2024 report)

individuals reporting life satisfaction; (2) indicators of a higher quality of social life—e.g., participatory governance, greater trust between citizens and the citizenry and government, a high level of stakeholder satisfaction, and lower crime and corruption; (3) highly effective means of creating public value—i.e., linking personal value pursuits and collective value interests and the individual good with the good of all; and (4) highly effective social planning and, in addition, established performance measures for ensuring achieving human, social, and sustainable development. In this respect, the frontrunners have a high level of GDP (compared to the other countries in the world) but rank high in providing the citizens with other primary social goods.

Countries that score high in multiple dimensions have a long history of taking into consideration the relationship between the concept of integrality (as it relates to social planning), cultural identity and values, and the role of technology in being well-integrated in the fabric of existence. The countries that are the frontrunners in the integrative model are outstanding in coordinating economic prosperity, social welfare, and environmental protection. They are able to balance market competition with fairness, honesty, trust, and extremely low levels of corruption. This means that sustainable future planning is inclusive of "A worldview that is guided by intentions, motivations, and ethical and moral choices that are rooted in integral values that drive individual and collective actions (Dessein et al., 2015, 32). The integrative framework serves to transform the way natural scientists, technologists, and social scientists think about human, social, and sustainable development. Thus, the integrative framework provides essential moral and ethical checks and balances in the era of Industry 4.0, artificial intelligence, and robotics. The effectiveness of the integrative mode is its inclusiveness of the ethical principles and values prescribed by the early environmental movement, the Earth Charter, Our Common Heritage, and the current UN Sustainability Agenda. Consequently, the model is both prescriptive and diagnostic in that it aligns planning aimed at achieving the basic aims of social action with planning to achieve sustainability goals.

The immensity of the current environmental problem poses a challenge to the social sciences to develop a framework for reducing human fragmentation and nature-human dualism. From the earliest stages of the contemporary environmental movement, there was a call for a multidimensional, holistic, and ethical approach to resolving the problem of human fragmentation by making development inclusive of human, social, and sustainable development. From the outset, the Earth Charter, for example, proclaimed that environmental problems, human problems, and social problems are interrelated and thus cannot be solved by merely focusing on satisfying material needs. Only a transdisciplinary, holistic, ethical, and participatory approach can resolve the problems resulting from the gaps in planning that occur as a result of the prior focus on increased production, consumption, and disposal. Classical social theorists, ethicists, perennial philosophers, and political philosophers described the aim of social action as achieving the good "To which all right actions are a means—right conduct, and goodness itself, is found in being in harmony with nature (Cicero, 2004, 27; also see Confucius, 1869, 233 & 306–309).

Although the preferences for development may vary between countries and even socio-cultural contexts within countries, it is good to have a framework that represents a consensus on the ethical considerations and value orientation connected with simultaneously improving economic conditions and developing advances in technology and innovations while, as well, improving performance in multiple dimensions (Our Common Future, 1987, 43). Without a broader framework for improving the human experience, the scope of our knowledge of what is relevant to social action (epistemology) is actually fragmented, and our choice of action is inappropriate action for enriching the human experience (ethics). Therefore, the environmental movement has evolved from its early eco-justice roots to its current sustainability discourse to prescribe the best means of achieving good in multiple dimensions. It does so in terms that are best for individuals, for their collective lives, and what maximizes the benefits they enjoy from their relationship with the environment. In short, it prescribes a means of how to enjoy the highest quality of social life.

References

Ahmed, R., Al-Mohannadi, D., & Linke, P. (2021). Multi-objective resource integration for sustainable industrial clusters. *Journal of Cleaner Production, 316*, 1–12.

Beckerman, W. (1994). Sustainable development: Is it a useful concept? *Environmental Values, 3*(3), 191–209.

Boulding, K. (2011). The economics of the coming spaceship earth. In H. Jarrett (Ed.), *Environmental quality in a growing economy* (pp. 3–14). New York, Earthscan.

Brauch, H., & Spring, Ú. (2016). Sustainability transition and sustainable peace: Scientific and policy context, scientific concepts and dimensions. In H. Brauch, Ú. Spring, J. Grin, &

J. Scheffran (Eds.), *Handbook on sustainability transition and sustainable peace*. Springer International Publishing.

Brown, D. (2014). Norms for scientific claims made in the face of scientific uncertainty. In L. Westra & M. Vilela (Eds.), *The earth Charter, ecological integrity and social movements*. Routledge.

Callenbach, E. (1975). *Ecotopia*. Banyan Tree.

Cassirer, E. (1953). *Substance and function*. (Swabey, William. & Swabey, Marie. Trans. Dover.

Cicero. (2004). *On moral ends*. Cambridge University Press.

Confucius. (1869). *The life and teachings of Confucius*. N. Trubner & Company.

Dalal-Clayton, B., & Sadler, B. (Eds.). (2004). *Sustainability appraisal: A sourcebook and reference guide to international experience*. Routledge.

Dessein, J., Soini, K., Fairclough, G., & Horlings, L. (2015). Culture in, for and as sustainable development. In *Conclusions from the COST action IS1007 investigating cultural sustainability*. University of Jyväskylä.

Earth Charter. (2000). *The earth Charter initiative*. Earth Charter International Secretariat.

Engles, R. (2014). In L. Westra & M. Vilela (Eds.), *The earth Charter, ecological integrity and social movements*. Routledge.

Escobar, A., Osterweil, M., & Sharma, K. (2024). *Relationality*. Bloomsbury Publishing.

Hart, S., & Milstein, M. (2003). Creating sustainable value. *The Academy of Management Executive, 17*(2), 56–69.

Howard, M., & King, J. (2004). The rise of neo-liberalism in advanced capitalist economies: towards a materialist explanation. In P. Arestis & M. Sawyer (Eds.), *The Rise of the Market: Critical Essays on the Political Economy of Neo-liberalism* (pp. 38–73). Edward Elgar.

Ibn Ashur, M. (2006). *Treatise on Maqasid al-Shariʿah*. (El-Mesawi, Mohamed. Trans. The International Institute of Islamic Thought.

Jonas, H. (1984). *The imperative of responsibility*. The University of Chicago Press.

Kibert, C., Monroe, M., Anna, P., Plate, R., & Paul, T. (2012). *Working toward sustainability: Ethical decision making in a technological world*. John Wiley & Sons, Inc..

Konold, D., & Schwietring, T. (2021). The great discrepancy: Political action, sustainable development and ecological communication. *Politics and Governance, 9*(1), 131–140.

Merchant, C. (2005). *Radical ecology*. Routledge.

Merli, R., Preziosi, M., & Acampora, A. (2018). How do scholars approach the circular economy? A systematic literature review. *Journal of Clean Production, 178*(2018), 703–722.

Miller, L. (2020). Self-determination versus techno-economic determinism. *Eruditio: Journal of the World Academy of Art and Sciences, 2*(5), 87–99.

Miller, L. (2022). *The theory and practice of value creation*. Cambridge Scholars Publishers.

Moore, M. (1995). *Creating public value: Strategic Management in Government*. Harvard University Press.

Niklas, L. (1977). The differentiation of society. *Canadian Journal of Sociology., 2*(1), 29–53.

Niklas, L. (1989). *Ecological Communication*. University of Chicago Press.

Nussbaum, M., & Sen, A. (1993). *The quality of life*. Oxford University Press.

OECD. (2011). *How's life? Measuring well-being*. OECD Publishing.

Our Common Future. (1987). *The world commission on environment and development*. University Press.

Porter, M., & Kramer, M. (2011). Creating shared value. *Harvard Business Review., 89*(1/2), 62–77.

Prahalad, C. K., & Ramaswamy, V. (2004). Co-creation experiences: The next practice in value creation. *Journal of Interactive Marketing, 18*(3), 5–14.

Schumpeter, J. (2006). *History of economic analysis*. Routledge.

Smith, G. (2013). *Methodological individualism. The system of liberty*. Cambridge University Press.

Spretnak, C. (1990). Ecofeminism: Our roots and flowering. In I. Diamond & G. Orenstein (Eds.), *Reweaving the world: The emergence of ecofeminism*. Sierra Club.

Stiglitz, J., Sen, A., & Fitoussi, J.-P. (2009). *The commission on the measurement of economic performance and social progress*. Centre de recherche en économie de Sciences Po.

ul Haq, M. (2003). *Human development in South Asia 2002*. Oxford University Press.
UN. (2004). *Government's primary responsibility. Commission for Social development*. The United Nations.
UN. (2021). *Frequently asked questions on human rights and climate change*. United Nations Human Rights Office.
UN. (2024). *The sustainable development goals report 2024*. United Nations Publications.
UNESCO. (2012). *Education for sustainable development sourcebook*. The United Nations Education, Scientific, and Cultural Organization.
Vargo, S., & Akaka, M. (2012). Value co-creation and service systems (re)formation: A service ecosystems view. *Service Science, 4*(3), 207–217.
Vargo, S., & Lusch, R. (2017). Service-dominant logic 2025. *International Journal of Research in Marketing., 34*, 46–67.
Wheatley, M. (2001). Restoring Hope to the future through critical education of leaders. *South African Business Review, 5*(2), 1–3.
World Bank. (2022). *Bhutan country environmental analysis*. World Bank.
Yunus, M. (2008). Creating a world without poverty. *Global Urban Development, 4*(2), 1–19.
Zavestoski, S., & Weigert, A. (2018). Bringing forth an Ecotopian future. In M. Wallace & D. Carruthers (Eds.), *Perma/culture: Imagining alternatives in an age of crisis*. Routledge.

The manufacturer's authorised representative in the EU is Springer Nature Customer Service Centre GmbH, Europaplatz 3, 69115 Heidelberg, Germany. If you have any concerns regarding our products, please contact ProductSafety@springernature.com

Printed and bound by CPI Group (UK) Ltd, Croydon, CR0 4YY
01/10/2026
02225936-0005